OXFORD

UNIVERSITY PRESS

C000025824

Essential Physics

for Cambridge IGCSE®

2nd Edition

Jim Breithaupt

Viv Newman

Darren Forbes

Series Editor:
Lawrie Ryan

OXFORD
UNIVERSITY PRESS

Great Clarendon Street, Oxford, OX2 6DP, United Kingdom

Oxford University Press is a department of the University of Oxford. It furthers the University's objective of excellence in research, scholarship, and education by publishing worldwide. Oxford is a registered trade mark of Oxford University Press in the UK and in certain other countries

Text © Jim Breithaupt 2015; end-of-unit summary questions © Oxford University Press 2015

The moral rights of the authors have been asserted

First published in 2015

All rights reserved. No part of this publication may be reproduced, stored in a retrieval system, or transmitted, in any form or by any means, without the prior permission in writing of Oxford University Press, or as expressly permitted by law, by licence or under terms agreed with the appropriate reprographics rights organization. Enquiries concerning reproduction outside the scope of the above should be sent to the Rights Department, Oxford University Press, at the address above.

You must not circulate this work in any other form and you must impose this same condition on any acquirer

British Library Cataloguing in Publication Data
Data available

978-0-19-839926-1

10 9 8 7 6 5 4 3 2 1

Paper used in the production of this book is a natural, recyclable product made from wood grown in sustainable forests. The manufacturing process conforms to the environmental regulations of the country of origin.

Typeset by GreenGate Publishing Services, Tonbridge, Kent

Illustrations include artwork drawn by GreenGate Publishing Services

Printed in China by Golden Cup

Acknowledgements
The publishers would like to thank the following for permissions to use their photographs:

Cover image: © Joe McBride/CORBIS; p3: HANK MORGAN/ SCIENCE PHOTO LIBRARY; p4: Shutterstock / Mega Pixel; p4: Brosko / Shutterstock; p6: CODY IMAGES/SCIENCE PHOTO LIBRARY; p8: KAJ R. SVENSSON/SCIENCE PHOTO LIBRARY; p10: © ANDREW WONG/Reuters/Corbis; p12: Martyn Chillmaid; p18: Martyn Chillmaid; p20: COLIN CUTHBERT/SCIENCE PHOTO LIBRARY; p22: © Grant Heilman Photography / Alamy; p26: Alamy/BL Images Ltd; p28: Daniel Goodings / Shutterstock.com; p31: Creative Commons / Public Domain; p32: Conrado / Shutterstock.com; p42: © Race-Press.com/dpa/Corbis; p44: Popperfoto / Contributor; p46: Barrett&MacKay; p50: TRL LTD./SCIENCE PHOTO LIBRARY; p51: Dorling Kindersley; p51: DR JEREMY BURGESS/ SCIENCE PHOTO LIBRARY.; p52: RICHARD R. HANSEN/ SCIENCE PHOTO LIBRARY; p53: © Mike Powell/Corbis; p54: HEALTH PROTECTION AGENCY/SCIENCE PHOTO LIBRARY; p54: ROBERT BROOK/SCIENCE PHOTO LIBRARY; p56: ALEX BARTEL/SCIENCE PHOTO LIBRARY; p57: MIKKEL JUUL JENSEN/BONNIER PUBLICATIONS/ SCIENCE PHOTO LIBRARY;

p58: MARTIN BOND/SCIENCE PHOTO LIBRARY; p58: DAVID HAY JONES/SCIENCE PHOTO LIBRARY; p59: Y.DEROME/ PUBLIPHOTO DIFFUSION/SCIENCE PHOTO LIBRARY; p60: G. BRAD LEWIS/SCIENCE PHOTO LIBRARY; p61: STEVE ALLEN/ SCIENCE PHOTO LIBRARY; p62: MARK TURNBALL/SCIENCE PHOTO LIBRARY; p64: NASA/SCIENCE PHOTO LIBRARY; p65: MAXIMILIAN STOCK LTD/SCIENCE PHOTO LIBRARY; p68: DR MORLEY READ/SCIENCE PHOTO LIBRARY; p69: © Image Source/Corbis; p70: LOUISE MURRAY/SCIENCE PHOTO LIBRARY; p71: COLIN CUTHBERT/SCIENCE PHOTO LIBRARY; p72: © John Henshall / Alamy; p76: SHEILA TERRY/SCIENCE PHOTO LIBRARY; p78: ANDREW LAMBERT PHOTOGRAPHY/ SCIENCE PHOTO LIBRARY; p79: CLIVE FREEMAN/BIOSYM TECHNOLOGIES/SCIENCE PHOTO LIBRARY; p83: SAM OGDEN/SCIENCE PHOTO LIBRARY; p88: DAVID TAYLOR/ SCIENCE PHOTO LIBRARY; p89: MARK BURNETT/SCIENCE PHOTO LIBRARY; p100: © Mark Boulton / Alamy; p102: JIM ZIPP/SCIENCE PHOTO LIBRARY; p102: © eye35.com / Alamy; p105: © Steve Bloom Images / Alamy; p106: CORDELIA MOLLOY/SCIENCE PHOTO LIBRARY; p106: SHEILA TERRY/ SCIENCE PHOTO LIBRARY; p108: MATT MEADOWS/SCIENCE PHOTO LIBRARY; p116: JPL CALTECH/STScI/VASSAR/NASA/ SCIENCE PHOTO LIBRARY; p120: ANDREW LAMBERT PHOTOGRAPHY/SCIENCE PHOTO LIBRARY; p120: © David R. Frazier Photolibrary, Inc. / Alamy; p122: VOLKER STEGER/ SCIENCE PHOTO LIBRARY; p124: ADAM HART-DAVIS/ SCIENCE PHOTO LIBRARY; p126: Jim Breithaupt; p129: DEEP LIGHT PRODUCTIONS/SCIENCE PHOTO LIBRARY; p130: TON KINSBERGEN/SCIENCE PHOTO LIBRARY; p130: IAN HOOTON/ SCIENCE PHOTO LIBRARY; p133: MAURO FERMARIELLO/ SCIENCE PHOTO LIBRARY; p134: DAVID PARKER/SCIENCE PHOTO LIBRARY; p136: © Medical-on-Line / Alamy; p137: © Corbis Premium RF / Alamy; p140: DARWIN DALE/ SCIENCE PHOTO LIBRARY; p144: KEITH KENT/SCIENCE PHOTO LIBRARY; p148: CORDELIA MOLLOY/SCIENCE PHOTO LIBRARY; p152: ADRIENNE HART-DAVIS/SCIENCE PHOTO LIBRARY; p157: ALEX BARTEL/SCIENCE PHOTO LIBRARY; p162: PETER MENZEL/SCIENCE PHOTO LIBRARY; p163: RIA NOVOSTI/SCIENCE PHOTO LIBRARY; p170: ANDREW LAMBERT PHOTOGRAPHY/SCIENCE PHOTO LIBRARY; p170: PATRICK DUMAS/EURELIOS/SCIENCE PHOTO LIBRARY; p172: ANDREW LAMBERT PHOTOGRAPHY/SCIENCE PHOTO LIBRARY; p178: JIM DOWDALLS/SCIENCE PHOTO LIBRARY; p200: © Leslie Garland Picture Library / Alamy; p201: Science Photo Library; p202: ANDREW LAMBERT PHOTOGRAPHY/ SCIENCE PHOTO LIBRARY; p203: Sandy Marshall; p207: GEOFF TOMPKINSON/SCIENCE PHOTO LIBRARY; p222: RIA NOVOSTI/SCIENCE PHOTO LIBRARY; p227: MCGILL UNIVERSITY, RUTHERFORD MUSEUM/EMILIO SEGRE VISUAL ARCHIVES/AMERICAN INSTITUTE OF PHYSICS/SCIENCE PHOTO LIBRARY; p231: MARTYN F. CHILLMAID/SCIENCE PHOTO LIBRARY; p233: Nelson Thornes.

Artwork by GreenGate Publishing Services and OUP .

Although we have made every effort to trace and contact all copyright holders before publication this has not been possible in all cases. If notified, the publisher will rectify any errors or omissions at the earliest opportunity.

All questions, example answers, marks awarded and comments that appear in this book were written by the author. In examination, the way marks are awarded to answers like these may be different. Cambridge International Examinations bears no responsibility for the example answers which are contained in this publication.

® IGCSE is the registered trademark of Cambridge International Examinations

Contents

Contents

Activities on the website

Test Yourself – Multiple choice tests for each Unit of the student book

On Your Marks – Activities which give guidance on how to approach exam questions by showing sample answers and teacher comments

Practical Physics – Interactive tests that focus on apparatus, taking measurements and best practice when answering questions that involve experiments and investigations

www.oxfordsecondary.com/
9780198399261

This book is designed specifically for Cambridge IGCSE® Physics 0625. Experienced teachers have been involved in all aspects of the book, including detailed planning to ensure that the content gives the best match possible to the syllabus.

Using this book will ensure that you are well prepared for studies beyond the IGCSE level in pure sciences, in applied sciences or in science-dependent vocational courses. The features of the book outlined below are designed to make learning as interesting and effective as possible:

STUDY TIPS

Experienced teachers give you suggestions on how to avoid common errors or give useful advice on how to tackle questions.

LEARNING OUTCOMES

- These are at the start of each spread and will tell you what you should be able to do at the end of the spread

Supplement

- Some outcomes will be needed only if you are taking a supplement paper and these are clearly labelled, as is any content in the spread that goes beyond the syllabus.

KEY POINTS

These summarise the most important things to learn from the spread.

SUMMARY QUESTIONS

These questions are at the end of each spread and allow you to test your understanding of the work covered in the spread.

At the end of each unit there is a double page of summary questions and examination-style questions written by the authors.

On the support website you will also find:

'Alternative to Practical' section – this provides guidance if you are doing this examination paper instead of coursework or the practical examination.

A useful Revision Checklist to help you prepare for the examination in four written papers.

A planning exercise example and guidance on how to write one.

DID YOU KNOW?

These are not needed in the examination but are found throughout the book to stimulate your interest in physics.

PRACTICAL

These show the opportunities for practical work. The results are included to help you if you do not actually tackle the experiment or are studying at home.

Assessment structure

Paper 1: Multiple Choice (Core)

Paper 2: Multiple Choice (Extended)

Paper 3: Theory (Core)

Paper 4: Theory (Extended)

Paper 5: Practical Test

Paper 6: Alternative to Practical

End of Unit answers are provided at

www.oxfordsecondary.com/ 9780198399261

Making measurements

Supplement

LEARNING OUTCOMES

- Use a metre rule to measure length accurately
- Use a clock to measure a time interval
- Use a measuring cylinder to measure liquid volume accurately
- Use and describe the use of a stopwatch to measure the period of oscillation of a pendulum
- Use and describe the use of a micrometer to measure short lengths accurately

SI units

quantity	unit	unit symbol
length	metre	m
time	second	s
mass	kilogram	kg
electric current	ampere	A
temperature	kelvin	K

Measurements at work

Many people need to make accurate measurements in their jobs. For example:

- a builder needs to measure lengths and distances accurately; stairs that are too short would be dangerous to climb and a door that doesn't fit exactly into the doorway would jam
- an air traffic controller needs to know the distance to an aircraft and how far it is above the ground as well as its estimated time of arrival
- a doctor or nurse giving an injection to a patient needs to fill the syringe with exactly the correct volume of the liquid to be injected.

A measurement has to be made accurately using a reliable instrument calibrated in agreed units. A reliable instrument must give the same reading every time the same quantity is measured. It must also pass the test of giving the correct reading when it is checked using a known quantity. In addition, it must be used correctly to obtain an accurate measurement. As we shall see below, even a simple metre rule or a stopwatch can be used inaccurately.

The units for scientific measurements are internationally agreed and are referred to as **SI units** (Systeme International). All scientific units you will meet in this course are derived from the five base units that are listed in the table on the left. Each unit is defined in terms of an agreed standard. For example, the unit of **mass** is the **kilogram** and is defined by a standard mass of 1 kilogram in the Bureau of International Weights and Measures in Paris.

Measuring length

We can use a **metre** rule graduated in millimetres to measure lengths to within a millimetre. For example, suppose we measure the **length** of a bar. The bar is placed alongside the rule with one end at the zero mark of the rule. Reading the position of the other end of the bar on the millimetre scale gives the length of the bar. However, common errors in such a measurement include:

- inaccurate setting of the end of the bar at the zero mark of the rule
- movement of the bar after setting one end at the zero mark
- 'line of sight' errors, as shown in Figure 1.1.1, when observing each end of the bar against the scale. The observer must be positioned so that the line of sight is at right angles to the scale.

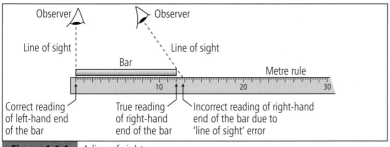

Figure 1.1.1   A line of sight error

Supplement

Using a micrometer

A micrometer is used to measure lengths of up to 30 mm to within 0.01 mm, as shown by the digital micrometer in use in the photo. A digital micrometer gives a read-out equal to the width of the micrometer gap directly. An analogue micrometer has a barrel on a screw thread with a pitch of 0.5 mm. For an analogue micrometer:

- the edge of the barrel is marked in 50 equal intervals so each interval corresponds to changing the gap of the micrometer by $\dfrac{0.5}{50}$ mm = 0.01 mm
- the stem of the micrometer is marked with a linear scale graduated in 0.5 mm marks
- the reading of a micrometer is where the linear scale intersects the scale on the barrel.

The analogue micrometer in Figure 1.1.2 shows a reading of 4.06 mm. The edge of the barrel is between the 4.0 and 4.5 mm marks on the linear scale. The linear scale intersects the sixth mark after the zero mark on the barrel scale. The reading is therefore 4.00 mm from the linear scale +0.06 mm from the barrel scale.

To use a micrometer correctly:

1 Check its zero reading and note the zero error if there is one.
2 Place the object in the gap then close the gap on the object to be measured. For an analogue micrometer, this involves turning the knob until it slips. Do not over-tighten the barrel.
3 Take the reading and note the measurement after allowing, if necessary, for the zero error.

An engineer using a digital micrometer

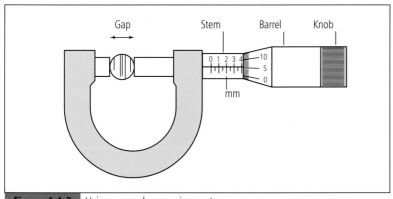

Figure 1.1.2   Using an analogue micrometer

KEY POINTS

The average thickness of two or more flat objects, e.g. coins, or sheets of paper, can be found by measuring their total thickness. Dividing this measurement by the number of objects gives their average thickness.

CONTINUED

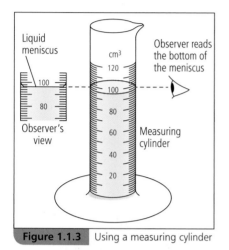

Figure 1.1.3 Using a measuring cylinder

Figure 1.1.5 Analogue and digital timers. The digital timer (top) gives a read-out of the timing. The timer below it is an example of an analogue measuring device as it has a pointer that must be read against the scale

Measuring volume

The volume of a liquid can be measured using a measuring cylinder. The scale is usually graduated in cubic centimetres (cm³). The liquid is poured into the empty measuring cylinder. Then the level of the liquid in the measuring cylinder can be read on the scale, as shown in Figure 1.1.3. The meniscus of the liquid surface is where it is in contact with the glass surface. The liquid level is at the bottom of the curved meniscus and the line of sight from the observer must be along this level.

Note

1 1 cubic metre (m³) = 1 000 000 cm³

2 The volume of an irregular solid can also be measured using a measuring cylinder. See topic 2.2.

Measuring time intervals

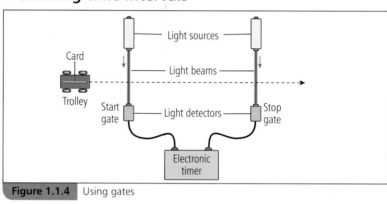

Figure 1.1.4 Using gates

To measure a time interval, we need to use an appropriate timing device. Such devices may be mechanical or electronic and they include ordinary clocks and watches, stopclocks, stopwatches and electronic light-gate timers. Examples of their use are listed in this table.

timing device	example of use
ordinary clock or watch	pay-to-park car parking
stopclock or stopwatch	timing oscillations, such as a swinging pendulum
electronic light-gate timer	100 m race

Timing devices need to be checked periodically to ensure they do not run too slow or too fast. Assuming a timing device is accurate, common errors in its use include:

• line of sight errors in reading the position of a pointer against the dial (if the line of sight from the observer to the pointer is not at right angles to the dial)

- reaction time errors when using a stopwatch or stopclock; such errors can be reduced with practice to about 0.1 s. Even so, timings of the same event by different people would be likely to differ by at least 0.1 s
- misalignment of the light beams when using a light-gate timer (if the light beams at the 'start' gate and the 'stop' gate are not parallel to each other).

Measurement of the period of a pendulum

A pendulum oscillates repeatedly with a constant time period. This is the time interval between successive passes in the same direction through the centre of the oscillations (i.e. its position when at rest). To measure its period:

- place a reference marker directly under the pendulum's rest position and observe the pendulum as it repeatedly moves past the marker
- ensure the line of sight to the marker is at right angles to the plane in which the pendulum oscillates
- time how long the pendulum takes to undergo a measured number of at least 10 oscillations from when it passes the marker in a certain direction
- repeat the procedure at least two more times and calculate the average time for the measured number of oscillations.

To calculate the time period, divide the average time for the measured number of oscillations by the number of oscillations.

STUDY TIP

Next time you are in a 100 m sprint race, make sure all the runners start in line and the finishing line is exactly at right angles to the track.

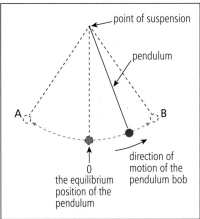

Figure 1.1.6 The time period of the oscillations is the time taken for the moving pendulum to travel from 0 to B, then down past 0 and up to A, and finally back to 0.

SUMMARY QUESTIONS

1 Copy and complete the following sentences using words from the list:

measuring cylinder   metre rule   micrometer millimetre scale

a A _____ with a _____ is used to measure lengths to within a millimetre.

b A _____ is used to measure volume.

c A _____ should be used to measure the diameter of a wire.

2 Explain how you would use a stopwatch to measure someone's pulse rate in beats per minute.

3 A student used a digital stopwatch three times to time 10 oscillations of a pendulum. The timings were 13.52 s, 13.64 s and 13.58 s.

a Calculate i the average time for 10 oscillations, ii the time period of the oscillations.

b Estimate the accuracy of the timings, giving a reason for your estimate.

KEY POINTS

1 A metre rule graduated in millimetres may be used to measure lengths that are too long to measure with a micrometer.

2 A measuring cylinder is used to measure liquid volume.

3 A stopwatch or stopclock is used to time oscillations.

4 A micrometer is used to measure very short distances accurately.

Supplement

Distance–time graphs

LEARNING OUTCOMES

- Interpret a distance–time graph to tell if an object is stationary
- Interpret a distance–time graph to tell if an object is moving at constant speed
- Calculate the speed of a body

Capturing the land speed record and becoming the first land vehicle to break the sound barrier

DID YOU KNOW?

- A top sprinter can travel a distance of about 10 m every second.
- A cheetah is faster than any other animal. It can run about 30 m every second – but only for about 20 s!
- A vehicle travelling at the speed limit of 108 kilometres per hour (km/h) on a motorway travels a distance of 30 m every second.
- The land speed record at present is 340 m/s, which is just over Mach 1, the speed of sound in air.
- The air speed record was broken in 2004 by X-43A, an experimental scram-jet plane. It reached Mach 9.6!

On a car journey

Some roads have marker posts every kilometre. If you are a passenger in a car on such a road, you can use these posts to check the speed of the car. You need to time the car as it passes each post. This table shows some measurements made on a car journey.

Data for distance–time graph

distance / metres (m)	0	1000	2000	3000	4000	5000	6000
time / seconds (s)	0	40	80	120	160	200	240

Figure 1.2.1 shows the readings on a graph of distance plotted against time.

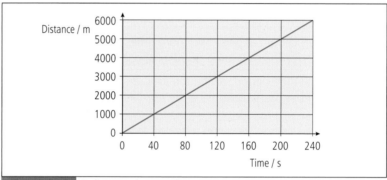

Figure 1.2.1 A distance–time graph

The graph shows that:

- the car took 40 s to go from each marker post to the next and so its speed was **constant**.
- the car travelled a distance of 25 m every second (= 1000 m ÷ 40 s) and so its speed was 25 m/s.

If the car had travelled faster, it would have gone further than 1000 m every 40 s and so the graph would have been steeper. In other words, the gradient of the graph would have been greater.

The gradient on a distance–time graph represents speed

- If the graph is flat, the gradient is zero and therefore the speed is zero.
- If the graph is straight and not flat, the gradient of the graph is constant and not zero. Therefore the speed is constant.
- If the graph is curved, the gradient of the graph changes and therefore the speed changes.

The steeper a distance–time graph is, the greater the speed it represents.

Speed

For an object moving at constant speed, we can calculate its speed using the equation:

$$\text{speed} = \frac{\text{total distance travelled}}{\text{total time taken}}$$

The scientific unit of speed is the metre per second, usually written as metre/second or m/s.

Note: for an object moving at changing speed, the equation above gives its **average speed**.

Long-distance journeys

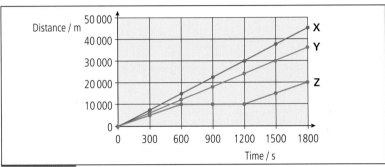

Figure 1.2.2 Comparing distance–time graphs

Long-distance vehicles are fitted with recorders that are used to check that drivers do not drive for too long. The information from a recorder may be used to plot a distance–time graph. Figure 1.2.2 shows distance–time graphs for three lorries: X, Y and Z on the same route.

- X went fastest because it travelled furthest in the same time.
- Y travelled more slowly than X. From the graph, you can check that it travelled 30 000 m in 1500 s. So its speed was 20 m/s (= 30 000 m ÷ 1500 s).
- Z stopped for some of the time. This is shown by the flat section of the graph. Its speed was zero during this time.

KEY POINTS

1 The steeper a distance–time graph is, the greater the speed it represents.

2 speed (metre/second; m/s) =

$$\frac{\text{distance travelled (metre; m)}}{\text{time taken (second; s)}}$$

SUMMARY QUESTIONS

1 Figure 1.2.3 shows the distance–time graphs for two model vehicles A and B plotted on the same axes. The two vehicles travelled the same distance.

   a Describe the motion of each vehicle in terms of its speed.

   b Calculate the average speed of each vehicle.

2 A vehicle on a road travels 1800 m in 60 s. Calculate:

   a the speed of the vehicle in m/s

   b how far it would travel at this speed in 300 s.

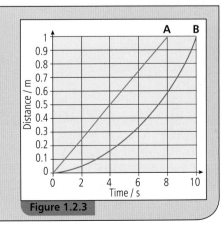

Figure 1.2.3

More about speed

Supplement

LEARNING OUTCOMES

- Interpret a distance–time graph to tell if an object's speed increases or decreases
- Measure the speed of an object
- Distinguish between speed and velocity

Using distance–time graphs

For an object moving at constant speed, we saw in Topic 1.2 that:

- the distance–time graph is a straight line
- the speed of the object is represented by the gradient of the line.

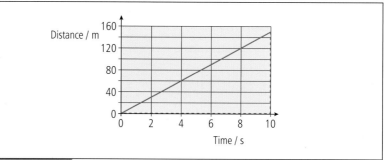

Figure 1.3.1 A distance–time graph for constant speed

To find the gradient of the line, we draw a triangle under the line, as in Figure 1.3.1. The height of the triangle represents the distance travelled and the base represents the time taken.

So the gradient of the line which represents the object's speed $=\dfrac{\text{the height of the triangle}}{\text{the base of the triangle}}$

For a moving object with changing speed, the distance–time graph is not a straight line. Figure 1.3.2 shows two examples. In both cases, the gradient of the line changes. This tells us that the speed of the object changes. For example, the gradient of the line in graph **a** increases because it becomes steeper as the time increases. Therefore the speed of the object that gave this line must be increasing. In other words, the object is accelerating.

Note When an object slows down, we say it **decelerates**.

A corkscrew ride

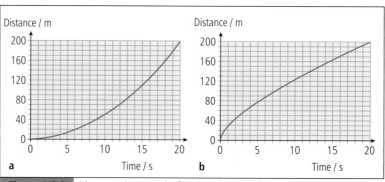

Figure 1.3.2 Distance–time graphs for changing speed

PRACTICAL

Measuring speed

1 Release a free-wheeling model vehicle or trolley at the top of a sloped runway. Adjust the runway so the trolley runs down as slowly as possible after being released at the top.

2 Use a stopwatch and a metre rule to make the measurements needed to plot a distance–time graph.

From your graph:

a Determine whether or not the speed of the trolley increased, decreased or remained constant as it moved down the runway.

b Calculate the average speed of the model vehicle over the first metre after it was released.

STUDY TIP

Next time you run around a track at a constant speed, think about why your velocity is changing even though your speed is constant.

Supplement

Speed and velocity

When you visit a fairground, the hardest rides are the ones that throw you around. Your speed and your direction of motion keep changing. We use the word **velocity** for speed in a given direction. An exciting ride would be one that changes your velocity often and unexpectedly!

Velocity is speed in a given direction

• An object moving at constant speed along a straight line has a constant velocity.

• An object moving steadily around in a circle has a constant speed. Its direction of motion changes as it goes around so its velocity is not constant.

• Two moving objects can have the same speed but different velocities. For example, a car travelling north at 30 m/s on a straight road has the same speed as a car travelling south at 30 m/s. But their velocities are not the same because they are moving in opposite directions.

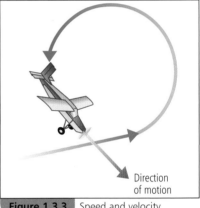

Direction of motion

Figure 1.3.3 Speed and velocity

SUMMARY QUESTIONS

1 Find the speed of the object graphed in Figure 1.3.1.

2 Describe how the speed of the object changes in Figure 1.3.2b.

Supplement

3 Copy and complete the following sentences below using words from the list:

acceleration   speed   velocity

a An object moving steadily around in a circle has a constant _____.

b If the velocity of an object increases by the same amount every second, its _____ is constant.

c When an object moves in a straight line with zero acceleration, its _____ is constant.

KEY POINTS

1 The gradient of the line on a distance–time graph gives the speed.

2 Velocity is speed in a given direction.

Supplement

Acceleration

Supplement

LEARNING OUTCOMES

- Interpret speed and time data to tell if an object is accelerating
- Interpret speed and time data to tell if the acceleration of an accelerating object is constant
- Use speed and time data to calculate the acceleration of an object undergoing constant acceleration

On a test track

On a test track

A car maker claims their new car 'accelerates more quickly than any other new car'. A rival car maker is not pleased by this claim and issues a challenge. Each car in turn is tested on a straight track with a speed recorder fitted. The results are shown in the table below. They show that the cars reach different speeds 6 seconds after the start which means that one accelerates more than the other.

time from a standing start / seconds (s)	0	2	4	6	8	10
speed of car X / metres/second (m/s)	0	5	10	15	20	25
speed of car Y / metres/second (m/s)	0	6	12	18	18	18

Which car accelerates more? The results are plotted on the speed–time graph in Figure 1.4.1. You can see that in the first 6 seconds, the speed of Y increases more than the speed of X. So Y accelerates more than X in the first 6 seconds.

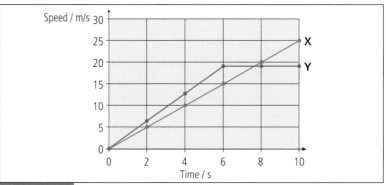

Figure 1.4.1 Speed–time graphs

Any object with an increasing speed is accelerating. Any object with a decreasing speed is decelerating. Figure 1.4.2 shows how the speed of car Y changed in the last 12 seconds of its journey on the test track. The graph shows that:

- the car moved at a constant speed of 18 m/s for the first 2 s
- the car's speed decreased from 18 m/s to 0 m/s in the next 10 s
- the car's speed was zero after 12 s.

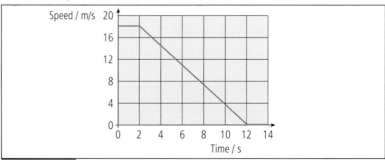

Figure 1.4.2 Decreasing speed

The acceleration of an object is its change of velocity per second. The unit of acceleration is the metre per second squared, abbreviated m/s².

We can work out the acceleration of an object using the equation:

$$\text{acceleration} = \frac{\text{change in velocity}}{\text{time taken for the change}}$$

In Figure 1.4.1, notice that:

- the velocity of X increases by 2.5 m/s every second. So the acceleration of X is constant and is equal to 2.5 m/s²
- the velocity of Y increases steadily for the first 6 s then remains constant. So Y's acceleration is constant for the first 6 s and then Y has zero acceleration from 6 s onward.

For an object moving with constant acceleration along a straight line, we can write the above word equation in symbols. Suppose the object's speed increases from an initial speed u to speed v in time t, as shown in Figure 1.4.3.

For motion along a straight line, acceleration, $a = \dfrac{\text{change of speed}}{\text{time taken}}$

$$= \frac{v - u}{t}$$

Multiplying both sides of this equation by t gives: $at = v - u$

Rearranging this equation gives:

$$v = u + at$$

Note An object moving at constant acceleration is sometimes said to have a **uniform** acceleration.

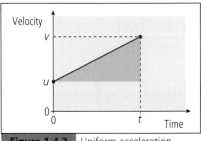

Figure 1.4.3 Uniform acceleration

SUMMARY QUESTIONS

1 The table below shows how the speed of a train changed between two stations.

time from start / seconds (s)	0	40	80	120	160	200	240
speed of train / metres/ second (m/s)	0	5	10	15	15	15	0

  a Use the data in the table to plot a graph of speed on the y-axis against time on the x-axis.
  b How long did the train spend **i** moving at constant speed, **ii** accelerating, **iii** decelerating?

  c Calculate the initial acceleration of the train.

2 The velocity of a car increased from 8 m/s to 28 m/s in 8 s without changing direction. Calculate:
  a its change of velocity and **b** its acceleration.

3 A jet plane accelerated along a straight runway from rest to a speed of 140 m/s in 50 s when it took off. Calculate:
  a its change of velocity and **b** its acceleration.

WORKED EXAMPLE

A motorcycle accelerates from rest for 15 s at a constant acceleration of 2.2 m/s². Calculate its velocity at the end of this time.

Solution

Initial velocity $u = 0$, acceleration $a = 2.2$ m/s².

At time $t = 15$ s,
velocity $v = u + at$
$= 0 + 2.2 \times 15 = 33$ m/s.

KEY POINTS

1 Acceleration is change of velocity per second.

2 The unit of acceleration is the metre/second² (m/s²).

More about acceleration

LEARNING OUTCOMES

- Recognise that the gradient of a speed–time graph tells us about acceleration

- Recognise that the area under a speed–time graph gives distance travelled

Supplement

- Use a speed–time graph to calculate the acceleration of an object which has a constant acceleration

Investigating acceleration

We can use a motion sensor linked to a computer to record how the speed of an object travelling along a straight line changes. Figure 1.5.1 shows how we can do this using a trolley as the moving object. The computer can also be used to display the measurements as a speed–time graph.

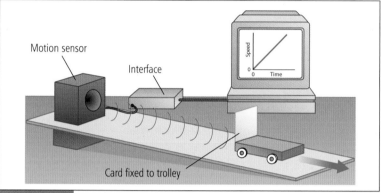

Figure 1.5.1   Investigating acceleration

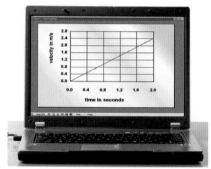

Measuring motion using a computer

Test A If we let the trolley accelerate down the runway, its speed increases with time. The computer on the left shows the speed–time graph from a test run.

- The line goes up because the speed increases with time. So it shows the trolley was accelerating as it ran down the runway.

- The line is straight which tells us that the increase of speed was the same every second. In other words, the acceleration of the trolley was constant (or *uniform*).

Test B If we make the runway steeper, the trolley accelerates faster. This would make the line on the graph steeper than for test A. So the acceleration in test B is greater.

The gradient of a graph is a measure of its steepness. The tests show that:

The gradient of a speed–time graph represents acceleration.

Changing acceleration

How is the acceleration affected in test A if the test is repeated with a 'windshield' attached to the front of the trolley. You should find that the gradient of the speed–time graph becomes less steep as the trolley gains speed, as shown by the graph in Figure 1.5.2. The graph shows that the acceleration is not constant and that it decreases as the speed increases. The air resistance due to the windshield causes this effect.

Figure 1.5.2   Decreasing acceleration

In the photograph on page 12:

- the gradient is given by the height divided by the base of the triangle under the graph
- the height of the triangle under the graph represents the change of velocity ($v - u$) and the base of the triangle represents the time taken t.

Therefore, the gradient represents the acceleration because

$$\text{acceleration} = \frac{\text{change of speed}}{\text{time taken}}$$

Prove for yourself that the acceleration in Figure 1.5.3 in the last 5 seconds is $-4.0\,\text{m/s}^2$. Note that this is a **negative acceleration** because it decelerates.

More about speed–time graphs

Figure 1.5.3 shows the speed–time graph for a vehicle braking to a standstill at a set of traffic lights. We use the term **deceleration** for any situation where an object decelerates.

- Before the brakes are applied, the vehicle moves at a constant speed of 20 m/s for 10 s. It therefore travels 200 m in this time ($= 20\,\text{m/s} \times 10\,\text{s}$). This distance is represented on the graph by the area under the graph from 0 to 10 s. This is the shaded rectangle in Figure 1.5.3.
- When the vehicle decelerates in Figure 1.5.3, its speed drops from 20 m/s to zero in 5 s. We can work out the distance travelled in this time from the area of the shaded triangle in Figure 1.5.3. This area is ½ × the height × the base of the triangle. So the vehicle must have travelled a distance of 50 m when it was decelerating.

The area under a speed–time graph represents distance travelled.

STUDY TIP

Next time you 'freewheel' on a bike down a slope, sit up and test the effect of air resistance on your speed.

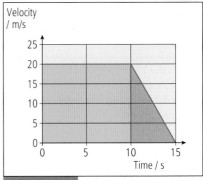

Figure 1.5.3 Braking

KEY POINTS

1 The gradient of a graph represents acceleration.

2 The area under a graph represents distance travelled.

SUMMARY QUESTIONS

1 Figure 1.5.4 shows four velocity–time graphs, labelled A, B, C and D.

  a Match each of the following descriptions to one of the graphs.

    i accelerated motion throughout

    ii zero acceleration

    iii accelerated motion then decelerated motion

    iv deceleration.

  b Which graph in Figure 1.5.4 represents the object that travelled:

    i the furthest distance

    ii the least distance?

2 Figure 1.5.5 shows the velocity–time graph of an object X moving with a constant acceleration.

  a How can you tell from the graph that the acceleration of X is constant?

  b Use the graph to calculate the distance moved by X in 10 s.

  c Determine the acceleration of the object.

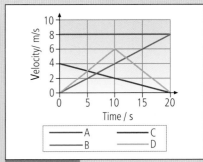

Figure 1.5.4

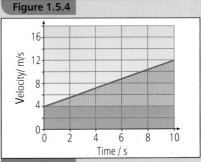

Figure 1.5.5

Free fall

Supplement

LEARNING OUTCOMES

- Recognise that objects in free fall accelerate at constant acceleration
- Know that the air resistance on objects in free fall is negligible
- Recognise that where air resistance is not negligible, a falling object reaches a terminal speed

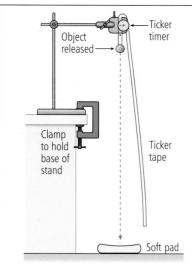

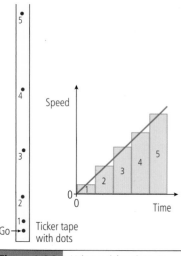

Figure 1.6.2 Using a ticker timer to show the acceleration of a falling object

Investigating free fall

Does a falling object gain speed as it falls? We can investigate the motion of a falling ball by different methods, two of which are described below.

Using a camera and a light flashing at a constant rate, we can take a 'multiflash' photograph of a falling ball, as shown in Figure 1.6.1. With a suitable light source flashing at a constant rate, each time the light flashes, an image of the falling object is recorded by the camera. The photograph shows that the distance between successive images of the ball increases as it falls. This means that the ball's speed increases as it falls.

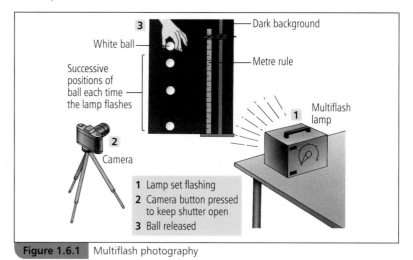

Figure 1.6.1 Multiflash photography

Using a ticker timer, as shown in Figure 1.6.2, we can make a tapechart to show the acceleration of a falling object is constant.

A paper tape attached to a suitable object is pulled through a ticker timer when the object is released. The ticker timer prints dots on the tape at a constant rate as the tape passes through it. The tape is then cut into single-dot lengths which are then stuck side-by-side on a sheet of paper to make a tape chart as shown in Figure 1.6.2.

- Each single-dot length is a measure of the speed of the object as the single-dot length passed through the ticker timer.
- The line through the tops of the tape lengths shows how the speed of the object changed as the object fell.
- The line on the tape chart in Figure 1.6.2 has a constant gradient. This shows that the speed of the falling object increased steadily. In other words, the object fell at constant acceleration.
- Any object released near the Earth's surface falls at a constant acceleration, provided air resistance is insignificant. Such motion is described as **free fall**.

The acceleration of a freely falling body near to the Earth's surface is constant.

Acceleration of free fall, *g*

The two methods described on page 14 can be used to measure the distance fallen by a falling object in different measured times. Such measurements may be used to show that a freely falling object has an acceleration of approximately 10 m/s². This acceleration is due to gravity and is referred to as the **acceleration of free fall, *g*.**

The Earth's gravitational field decreases in strength with increased height above the Earth's surface. However, for heights which are very small compared with the Earth's radius, the field is effectively uniform (i.e. the same everywhere) with a value of *g* equal to approximately 10 m/s².

The effect of air resistance on falling objects

A parachutist who jumps out of a plane accelerates until the parachute opens. The parachutist descends to the ground at constant speed because the air resistance on the parachute (and the parachutist) opposes the force of gravity on them (i.e. their total weight) with an equal force. The air resistance is sometimes referred to as the **drag force**.

In general, the air resistance on any falling object increases as it gains speed. If it continues to fall, the increasing air resistance causes it to reach a constant speed vertically downwards, referred to as its **terminal velocity**. At this speed, the air resistance on the object opposes the force of gravity with an equal force. Figure 1.6.4 shows how the speed of such a falling object increases as it descends until it reaches its terminal speed.

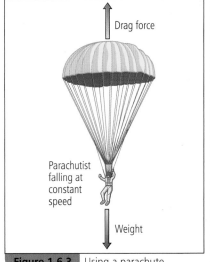

Figure 1.6.3   Using a parachute

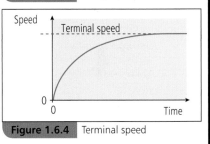

Figure 1.6.4   Terminal speed

SUMMARY QUESTIONS

1 Two objects, X and Y, are released simultaneously from different heights above the ground. X is released above Y, as shown in Figure 1.6.5.

Copy and complete the following sentences below using words from the list.

a greater   the same   a smaller

a Compared with X, Y has _____ acceleration as they fall.

b Compared with X, Y hits the ground with _____ speed.

c Compared with X, Y has _____ time of descent.

2 a Which feature in the graph in Figure 1.6.4 represents:

   i   the distance fallen after the object was released?

   ii   the acceleration of the object?

b Describe how the acceleration of the object changed after it was released.

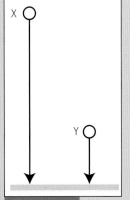

Figure 1.6.5

KEY POINTS

1 A freely falling object has a constant acceleration.

2 The acceleration of a freely falling object near the Earth is 10 m/s².

3 If air resistance is significant, a falling object reaches a terminal speed.

1 (a) Explain, with the aid of a diagram, how to avoid a 'line of sight' (or parallax) error when taking a measuring cylinder reading.

(b) The liquid surface in the measuring cylinder will not be entirely flat. Draw a diagram to show the shape of the meniscus and to show where to take the reading.

2 When using a pendulum, the experiment usually involves finding the period of the pendulum.

(a) Explain the meaning of the term 'period'. You may draw a diagram to help your description.

(b) A careful experimenter will measure the time for at least ten oscillations of the pendulum and then calculate the period. Explain why it is good practice to measure the time for at least ten oscillations.

3 The following sketch graphs show the motion of a toy car:

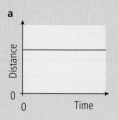
a

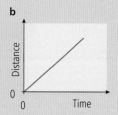
b

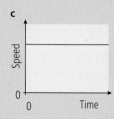
c

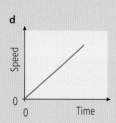
d

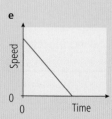
e

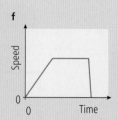
f

State briefly what each graph tells you about the motion.

The speed–time graph shows the movement of a car.

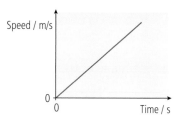

1 The graph shows that the car is

  A accelerating

  B moving at a constant speed

  C moving up a hill

  D stationary

(Paper 1/2)

2 The gradient of a distance–time graph represents

  A acceleration

  B distance travelled

  C speed

  D time taken

(Paper 1/2)

3 The area under a speed–time graph represents

  A acceleration   **C** speed

  B distance travelled   **D** time taken

(Paper 1/2)

4 The table shows the results obtained by a student measuring the speed of a trolley travelling down a ramp.

time / s	speed / m/s
0	0
0.2	1.4
0.4	3.0
0.6	4.6
0.8	6.1
1.0	7.4
1.2	9.0

(a) Use the information in the table to draw a speed–time graph of the motion of the trolley.

(b) Use the graph to determine the acceleration of the trolley.

Supplement

4

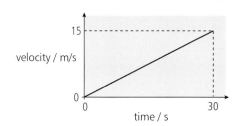

The graph shows the motion of a vehicle. The acceleration, in m/s², of the vehicle is

A 0.5 **B** 2 **C** 15 **D** 225 [1]

(Paper 1/2)

5

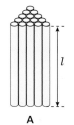

A   B   C

Diagram A shows a bundle of wooden rods. They are shown actual size.

(a) On diagram A measure the length l of a rod. [2]

(b) (i) On diagram B measure x, the combined diameters of five rods.

  (ii) Calculate the diameter d of one rod. [2]

(c) Explain briefly why it is good experimental practice to measure x to find the diameter. [1]

(d) Calculate the volume V of one rod using the equation $V = \dfrac{\pi d^2 l}{4}$ [2]

(e) The rods are closely packed into a thin cardboard box as shown in diagram C.

  (i) Estimate the volume of the box.

  (ii) Explain briefly how you arrived at your estimated value. [2]

(Paper 6)

6 The table shows the readings obtained by a student during an experiment in which a toy car runs down a slope.

Time / s	Distance / m
0	0
0.2	0.50
0.4	0.95
0.6	1.35
0.8	1.70
1.0	2.00
1.2	2.25
1.4	2.50
1.6	2.75
1.8	2.90
2.0	2.90

(a) Draw a distance–time graph using the readings in the table. [4]

(b) State what the graph tells you about the motion of the toy car:

  (i) during the first second

  (ii) from 1.0 to 1.6 seconds

  (iii) from 1.8 to 2.0 seconds. [4]

(Paper 3)

7 In an experiment to determine the acceleration of free fall, g, a student attached a length of ticker tape to a metal ball. When the ball was released it fell to the ground pulling the tape through a ticker timer that printed 50 dots per second on the tape. The student measured lengths of tape to find the speed of the ball as it fell.

(a) Calculate the time taken for the timer to print 10 dots on the tape. [2]

(b) The student cut the tape into lengths with 10 dots each, starting at the end attached to the ball. Would you expect the lengths to increase, decrease or stay the same as the ball fell? [1]

(c) The student's results showed that after 0.6 s the speed of the ball was 5.2 m/s. Calculate a value for the acceleration of free fall, g, using the student's results. [3]

(d) The value obtained in part **(c)** is less than the accepted value. Suggest a reason for this difference, assuming that the experiment was carried out with care. [1]

(Paper 4)

Supplement

LEARNING OUTCOMES

- Recognise that the mass of a body is a measure of how much matter is in it
- Compare different masses using a balance
- Recognise that the weight of an object depends on its mass
- Recognise that the weight of an object depends on the gravitational field it is in
- Know that the greater the mass of an object is, the greater the resistance to change of its motion

Using kilograms

PRACTICAL

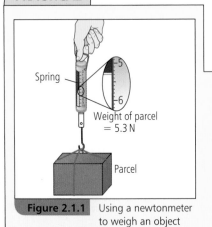

Spring

Weight of parcel = 5.3 N

Parcel

Figure 2.1.1 Using a newtonmeter to weigh an object

Mass and matter

The mass of an object depends on how much matter there is in the object. The amount of matter in an object determines its **mass,** regardless of whether the object is a solid or a liquid or a gas. Two objects of the same mass contain the same amount of matter. Two objects of different mass contain different amounts of matter.

The SI unit of mass is the **kilogram** (kg). We usually use this unit of mass in everyday life although we sometimes find it is more convenient to use the gram which is 0.001 kg.

The mass of a body is a measure of the amount of matter it contains.

Weight

The weight of an object depends on its mass. This is because weight is due to the downward pull of the Earth's gravity on the object and the force of gravity on an object depends on its mass.

The greater the mass of an object is, the greater its weight is.

We measure weight in newtons because the SI unit of force is the **newton** (abbreviated N) and weight is a force. Figure 2.1.1 shows an object being weighed using a **newtonmeter** marked in newtons. Measurements using a newtonmeter should show that the weight of an object of mass 1 kg near the Earth's surface is 10 N. Therefore the force of gravity on a 1 kg object near the surface of the Earth is 10 N.

For any object near the Earth's surface, the force of gravity on it is 10 N for every kilogram of its mass. So the weight of an object near the Earth's surface is 10 N for every kilogram of its mass. For example, near the surface of the Earth, the weight of an object:

- of mass 1 kg is 10 N
- of mass 5 kg is 50 N
- of mass 20 kg is 200 N.

Using a newtonmeter

1 Check the pointer of the newtonmeter reads zero on the scale without any object suspended from the newtonmeter.

2 Suspend the object to be weighed from the newtonmeter hook. This causes the spring in the newtonmeter to stretch which makes the pointer move down the scale.

3 Read the position of the pointer on the scale to give the weight of the object in newtons.

Gravitational field strength

The force of gravity on any object near the Earth's surface is 10 N for every kilogram of its mass. We say that the **gravitational field strength** of the Earth near its surface is 10 N/kg.

If we know the mass of an object, we can calculate its weight using the equation:

weight = mass × gravitational field strength
(in newtons) (in kilograms) (in N/kg)

The weight of an object depends on its location. For example, the weight of a 50 kg person near the Earth's surface is 500 N (= 50 kg × 10 N/kg). However, the same person on the surface of the Moon where the gravitational field strength is 1.6 N/kg would weigh only 80 N (= 50 kg × 1.6 N/kg).

Comparing masses

We can compare the masses of two different objects using a balance as shown in Figure 2.1.2. For two objects of the same mass, the arm of the balance would be level. For two objects of different mass, the end of the arm supporting the heavier one would drop and the other end would rise.

We can find the mass of an object by placing it on one of the balance pans and placing 'standards' of known mass on the other pan until the arm is level.

Inertia

The **inertia** of an object is its resistance to a change of its motion. The greater the mass of an object, the more inertia it has. A fully loaded lorry takes longer to reach a certain speed from a standstill than if it were carrying no load. Its mass when fully loaded is much greater than when it is unloaded, so it has more inertia and takes longer to accelerate from rest to a certain speed.

SUMMARY QUESTIONS

1 Copy and complete the following sentences below using words from the list.

force   matter
mass   weight

a The _____ of an object is a measure of how much _____ it has.

b The _____ of an object is the _____ on it due to gravity.

c _____ is measured in newtons; _____ is measured in kilograms.

2 An object has a mass of 40 kg on the surface of the Earth.

a State whether **i** its mass, **ii** its weight would be smaller, the same or greater on the Moon.

b Calculate **i** the weight of the object on the Earth, **ii** its weight on the Moon.

Use the gravitational field strengths given in the worked example.

WORKED EXAMPLE

Calculate the weight in newtons of a person of mass 60 kg:

a near the Earth's surface

b on the surface of the Moon.

The gravitational field strength near the Earth's surface = 10 N/kg.

The gravitational field strength near the Moon's surface = 1.6 N/kg.

Solution

a Near the Earth's surface, the weight of the person = mass × gravitational field strength
= 60 kg × 10 N/kg = 600 N.

b On the Moon's surface, the weight of the person = mass × gravitational field strength
= 60 kg × 1.6 N/kg = 96 N.

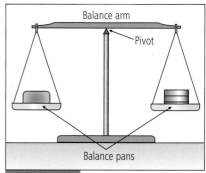

Figure 2.1.2   A balance to compare masses

KEY POINTS

1 The greater the mass of an object the greater is its weight.

2 The greater the mass of an object, the greater is its inertia.

Density

LEARNING OUTCOMES

- Recognise that density = mass / volume
- Calculate the density of an object from its mass and its volume
- Carry out and describe experiments to measure the density of a regular solid, of a liquid, and of an irregularly shaped solid
- Carry out calculations using the equation: 'density = mass / volume'

WORKED EXAMPLE

A wooden post has a volume of 0.025 m³ and a mass of 20 kg. Calculate its density in kg/m³.

Solution

$$density = \frac{mass}{volume} = \frac{20\,kg}{0.25\,m^3}$$

$$= 800\,kg/m^3$$

Materials of different densities

Density comparisons

Any builder knows that a concrete post is much heavier than a wooden post of the same size. This is because the density of concrete is much greater than the density of wood. A volume of one cubic metre of wood has a mass of about 800 kg whereas a cubic metre of concrete has a mass of about 2400 kg. So the density of concrete is about three times the density of wood.

The density of two different materials can be compared by comparing the masses of same-size blocks of each material. We can do this using a balance as shown in Figure 2.1.2 on page 19 or we can use an electronic balance to measure the mass of each block. Each block has the same volume so the block with the greater mass has the greater density.

The density (ρ) of a substance is defined as its mass (m) per unit volume (V). We can use the equation below to calculate the density of a substance if we know the mass and the volume of a sample of it.

$$density = \frac{mass}{volume} \quad or \quad \rho = \frac{m}{V}$$

The SI unit of density is the kilogram per cubic metre (kg/m³) although the gram per cubic centimetre (g/cm³) is often used.

Density tests

For each of the tests below, measure the mass and the volume of the object as explained then use the formula, $density = \frac{mass}{volume}$ to calculate the density of the object.

1 Measuring the density of a regular solid object

- To measure the mass of the object, use a balance as shown in Figure 2.1.2 on page 19 or an electronic balance. Make sure the balance reads zero before placing the object on it.

- To find the volume of a regular solid such as a cube, a cuboid or a cylinder, measure its dimensions, using a millimetre ruler or a micrometer. Use the measurements and the correct formula shown in Figure 2.2.1 to calculate its volume.

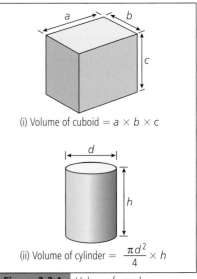

(i) Volume of cuboid = $a \times b \times c$

(ii) Volume of cylinder = $\frac{\pi d^2}{4} \times h$

Figure 2.2.1 Volume formulae

2 Measuring the density of a liquid

- Use a measuring cylinder to measure the volume of a certain amount of the liquid.
- Measure the mass of an empty beaker using a balance. Remove the beaker from the balance and pour the liquid from the measuring cylinder into the beaker. Use the balance again to measure the total mass of the beaker and the liquid. The mass of the liquid is worked out by subtracting the mass of the empty beaker from the total mass of the beaker and the liquid.

An object will float in a liquid if its density is less than the density of the liquid. You could test this using the results of your measurements.

WORKED EXAMPLE

A measuring cylinder contained a volume of 120 cm³ of a certain liquid. The liquid was then poured into an empty beaker of mass 51 g. The total mass of the beaker and the liquid was then found to be 145 g.

a Calculate the mass of the liquid in grams.

b Calculate the density of the liquid in g/cm³.

Solution

mass of liquid = 145 − 51 = 94 g; volume = 120 cm³

$$\text{density} = \frac{\text{mass}}{\text{volume}} = \frac{94\,\text{g}}{120\,\text{cm}^3} = 0.78\,\text{g/cm}^3$$

Measuring the density of an irregular solid

- Use a balance to measure the mass of the object.
- Determine the volume of the object using a beaker and a displacement can as shown in Figure 2.2.2. Water is the most suitable liquid to use provided the solid does not dissolve in it. Work out the density from the density equation below:

$$\text{density} = \frac{\text{mass}}{\text{volume}}$$

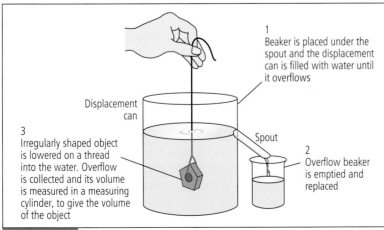

1 Beaker is placed under the spout and the displacement can is filled with water until it overflows

Displacement can

3 Irregularly shaped object is lowered on a thread into the water. Overflow is collected and its volume is measured in a measuring cylinder, to give the volume of the object

Spout

2 Overflow beaker is emptied and replaced

Figure 2.2.2 Measuring the volume of an irregular object in a measuring cylinder

SUMMARY QUESTIONS

1 A rectangular concrete slab is 0.80 m long, 0.60 m wide and 0.05 m thick.

  a Calculate its volume in m³.

  b The mass of the concrete slab is 60 kg. Calculate its density in kg/m³.

2 A measuring cylinder contains 80 cm³ of a certain liquid. The liquid is poured into an empty beaker of mass 48 g. The total mass of the beaker and the liquid was found to be 136 g.

  a Calculate the mass of the liquid in grams.

  b Calculate the density of the liquid in g/cm³.

3 A rectangular block of gold is 0.10 m in length, 0.08 m in width and 0.05 m in thickness.

  a **i** Calculate the volume of the block.

      ii If the mass of the block is 0.76 kg, calculate the density of gold.

  b A thin gold sheet has a length of 0.15 m and a width of 0.12 m. The mass of the sheet is 0.0015 kg. Use these measurements and the result of your density calculation in **a ii** to calculate the thickness of the sheet.

4 Describe how you would measure the density of a metal bolt. You may assume the bolt will fit into a measuring cylinder of capacity 100 cm³.

Supplement

LEARNING OUTCOMES

- Describe how to measure the extension of an object when it is stretched
- Recognise that an elastic body regains its shape after being deformed
- Describe extension–load graphs for a spring, for rubber and for polythene
- Interpret extension–load graphs including the limit of proportionality
- State and use Hooke's law

DID YOU KNOW?

Rubber and other soft materials dipped in liquid nitrogen become as brittle as glass. Such frozen materials shatter when struck with a hammer.

A shattered flower

Stretching and squeezing

Squash players know that hitting a squash ball changes its shape briefly. An object is said to be **elastic** if it regains its original shape when the forces that deform it are removed. A squash ball is elastic because it regains its shape. So too is a rubber band, as it regains its original length after it is stretched and then released. Rubber is an example of an elastic material.

Stretch tests

We can investigate how easily a material stretches by hanging weights from it, as shown in Figure 2.3.1.

- The strip of material under test is clamped at its upper end and its initial length is measured using the metre rule. A small weight or a weight hanger attached to the material is used to keep it straight.
- The amount of weight hung from the material is then increased in stages. The strip stretches each time more weight is hung from it.
- At each stage, the total weight added is recorded in a table and the length of the strip is measured and also recorded in the table. The position of the upper end should stay the same throughout.

The change of length from the initial length is called the **extension**. This is calculated for each stage and recorded in the table, as shown below.

The extension of the strip of material at any stage = its length at that stage – its initial length

weight / N	length / mm	extension / mm
0	120	0
1.0	152	32
2.0	190	70
3.0	250	130

Force versus length measurements for a rubber strip

The measurements may be plotted on a graph of extension on the vertical axis against weight on the horizontal axis. Figure 2.3.2 shows the results for strips of different materials and a steel spring plotted on the same axes.

- The steel spring gives a straight line through the origin. This shows that the extension of the steel spring is directly proportional to the weight suspended on it. For example, doubling the weight from 2.0 to 4.0 N doubles the extension of the spring.
- The rubber band does not give a straight line. When the weight on the rubber band is doubled from 2.0 to 4.0 N, the extension more than doubles.
- The polythene strip does not give a straight line either. As the weight is increased from zero, the polythene strip stretches very little at first then it 'gives' and stretches easily.

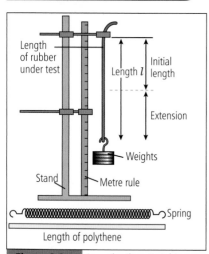

Length of rubber under test

Length *l* Initial length

Extension

Weights

Stand Metre rule

Spring

Length of polythene

Figure 2.3.1 Investigating stretching

Hooke's law for springs states that the extension of a spring is directly proportional to the weight it supports.

Notes

1   Hooke's law applies up to a limit known as **the limit of proportionality**. The graphs in Figure 2.3.2 show that rubber and polythene have a low limit of proportionality. A steel spring has a much higher limit of proportionality.

2   Hooke's law may be written as an equation

$$F = kx$$

where F is the stretching force or tension, x is the extension and k is the **spring constant**. A graph of F against x gives a straight line through the origin (Figure 2.3.3).

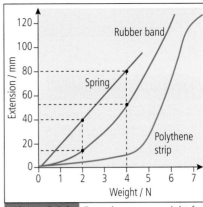

Figure 2.3.2   Extension versus weight for different materials

Supplement

KEY POINTS

1   The extension of a strip of stretched material is its extended length minus its initial length.

2   The extension of a spring is directly proportional to the weight it supports.

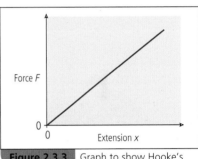

Figure 2.3.3   Graph to show Hooke's law

SUMMARY QUESTIONS

1   Copy and complete the following sentences below using words from the list.

extension   length

a   When a steel spring is stretched, its _____ is increased.

b   When a strip of polythene is stretched too much, its _____ is permanent.

c   When rubber is stretched and unstretched, its _____ afterwards is zero.

2   Describe how you would use the arrangement in Figure 2.3.1 to find out if a strip of material is an elastic material.

3   In a Hooke's law test on a spring, the following results were obtained.

weight / N	0	1.0	2.0	3.0	4.0	5.0	6.0
length / mm	245	285	324	366	405	446	484
extension / mm	0	40					

a   Complete the bottom row of the table.

b   Plot a graph of the extension on the vertical axis against the weight on the horizontal axis.

c   If a weight of 7.0 N is suspended on the spring, what would be the extension of the spring?

d   i   Calculate the spring constant of the spring.

ii   An object suspended on the spring gives an extension of 140 mm. Calculate the weight of the object.

Supplement

Force and motion

Supplement

LEARNING OUTCOMES

- Describe how a force may change the motion of an object
- Recognise that a resultant force acts on an object when the object accelerates or decelerates
- Recall and use the equation 'force = mass × acceleration'

Figure 2.4.1 Overcoming friction

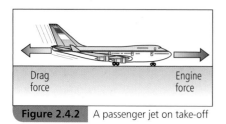

Figure 2.4.2 A passenger jet on take-off

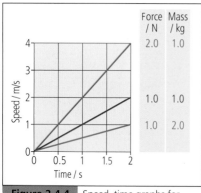

	Force / N	Mass / kg
	2.0	1.0
	1.0	1.0
	1.0	2.0

Figure 2.4.4 Speed–time graphs for different forces and masses

Most objects around us are acted on by more than one force. We can work out the effect of the forces on the motion of an object by replacing them with a single force, the **resultant force**. This is a single force that has the same effect as all the forces acting on the object.

When the resultant force on an object is zero, the object:

- remains at rest if it was already at rest
- continues to move at the same speed and in the same direction if it was already moving.

For example, when a heavy crate is pushed across a rough floor, the crate moves at constant speed across the floor. The push force on the crate is equal and opposite to the force of friction of the floor on the crate. The resultant force on the crate is therefore zero. Frictional forces oppose the motion of any two surfaces that slide (or try to slide) across each other.

When the resultant force on an object is not zero, the movement of the object depends on the size and direction of the resultant force.

For example, when a jet plane is taking off, the force of its engines is greater than the force of air resistance on it. Air resistance is a form of friction. The resultant force on it is the difference between the thrust force and the force of air resistance on it. The resultant force is therefore not zero. The greater the resultant force, the sooner the plane reaches its take-off speed.

PRACTICAL

Investigating force and acceleration

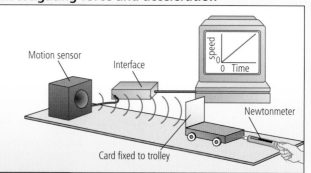

Figure 2.4.3 Investigating the link between force and motion

We can use the apparatus in Figure 2.4.3 to investigate how the acceleration of a trolley depends on the resultant force acting on it.

1 A newtonmeter is used to pull the trolley along with a constant force.

2 The total moving mass can be doubled or trebled by using double-deck and triple-deck trolleys.

3 A motion sensor and a computer are used to record the speed of the trolley as it accelerates. The results are displayed as a speed–time graph on the computer screen.

Figure 2.4.4 shows speed–time graphs for different amounts of force. The gradient of each line gives the acceleration. These show that

- **for a given mass, the greater the force, the greater the acceleration**
- **for a given force, the greater the mass, the smaller the acceleration.**

STUDY TIP

If an object is accelerating there must be a resultant force acting on it.

An equation for force and acceleration

We can work out the acceleration from the gradient of the graph, as explained in Topic 1.4.

Some typical results are given in the table below.

resultant force / N	0.5	1.0	1.5	2.0	4.0	6.0
mass / kg	1.0	1.0	1.0	2.0	2.0	2.0
acceleration / m/s^2	0.5	1.0	1.5	1.0	2.0	3.0
mass × acceleration / kg m/s^2	0.5	1.0	1.5	2.0	4.0	6.0

The results show that the resultant force, the mass and the acceleration are linked by the equation:

resultant force / N = mass / kg × acceleration / m/s^2

Mathematics notes

1 The word equation above may be written in the form:

resultant force $F = ma$, where m = mass and a = acceleration

2 Rearranging this equation gives $a = \dfrac{F}{m}$ or $m = \dfrac{F}{a}$

WORKED EXAMPLE

Calculate the resultant force on an object of mass 6.0 kg when it has an acceleration of 3.0 m/s^2.

Solution
Resultant force $F = ma = 6.0\,\text{kg} \times 3.0\,\text{m/s}^2 = 18.0\,\text{N}$.

KEY POINTS

	object at the start	resultant force	effect on the object
1	at rest	zero	stays at rest
2	moving	zero	speed and direction of motion stay the same
3	moving	non-zero in the same direction as the direction of motion of the object	accelerates
4	moving	non-zero in the opposite direction to the direction of motion of the object	decelerates

Resultant force (N) = mass (kg) × acceleration (m/s^2)

SUMMARY QUESTIONS

1 Copy and complete the sentences below using words from the list.

acceleration   motion   mass   resultant force   speed

a A moving object decelerates when a _____ acts on it in the opposite direction to its _____ .

b The greater the _____ of an object, the less its acceleration when a _____ acts on it.

c The _____ of a moving object increases when a _____ acts on it in the same direction as it is already moving.

2 A jet plane lands on a runway and stops.

a What can you say about the direction of the resultant force on the plane as it lands?

b What can you say about the resultant force on the plane when it has stopped?

3 Copy and complete the following table.

	force / N	mass / kg	acceleration / m/s^2
a	?	20	0.80
b	200	?	5.0
c	840	70	?
d	?	0.40	6.0
e	5000	?	0.20

Supplement

25

More about force and motion

Supplement

LEARNING OUTCOMES

- Recognise how to find the resultant of two forces that act along the same straight line
- Describe how to represent a force as a vector
- Recognise that an object in circular motion is acted on by a centripetal force that acts towards the centre of the circle

A train on an incline

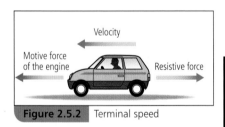

Figure 2.5.2 Terminal speed

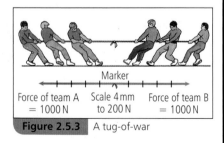

Figure 2.5.3 A tug-of-war

On the move

A railway engine pulling a train of carriages and wagons along a track needs to have enough power to pull the train up the steepest incline on the track. If the engine power is not enough, a second engine could be used to help. The force of the two engines on the train is equal to the sum of the force of each engine on the train. For example, if one engine pulls with a force of 10 000 N on the train and the other pushes with a force of 8000 N, the total force on the train is 18 000 N (= 10 000 N + 8000 N).

A car stuck in mud can be difficult to shift. A tractor can be very useful here. Figure 2.5.1 shows the idea. One end of the rope is tied to the back of the tractor and the other end to the front of the car. To pull the car out of the mud, the force of the tractor on the car needs to be greater than the force of the mud on the car. If the force of the mud on the car is equal to the force of the tractor on the car, the car stays stuck in the mud.

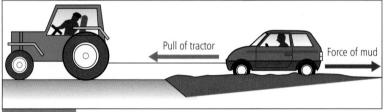

Figure 2.5.1 In the mud

A vehicle moving at its terminal speed on a flat road is pushed forward by the 'motive' force of its engine and opposed by a resistive force due to drag and friction. At terminal speed, the resistive force is equal and opposite to the engine force (Figure 2.5.2). Therefore, the resultant force is zero and so the acceleration is zero.

Supplement

Vectors

- The size and direction of a force can be represented by a **vector**. A vector is an arrow of length that represents the magnitude (size) of the force in the direction of the force. Any force has magnitude and direction and so can be drawn as a vector.

- Figure 2.5.3 shows the pull of two tug-of-war teams as vectors. A scale of 4 mm to 200 N is used here so the force of 1000 N is represented by a vector 2 cm long. In this example, the magnitudes of the two forces are the same so the two vectors are the same length. Because the forces are in opposite directions, the two vectors point in opposite directions.

- Suppose one team pulls with a force of 1000 N and the other team with a force of 750 N. The vector diagram for this situation is shown in Figure 2.5.4. The smaller force nearly cancels out the other force, but not quite. The stronger team exerts a force which is 250 N greater than the other team. So the resultant force (i.e. their combined effect) is 250 N.

In general, if an object is acted on by two forces:

- in the same direction, the resultant force is the sum of the two forces
- in opposite directions, the resultant force is the difference between the two forces.

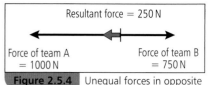

Figure 2.5.4 Unequal forces in opposite directions

Going around in circles

A satellite in a circular orbit above the Earth moves along its orbit at constant speed. The only force acting on the satellite is the force of gravity on it due to the Earth. This force acts towards the centre of the Earth, pulling on the satellite to prevent it from flying off 'at a tangent' into space. So it changes the direction of motion of the satellite without changing its speed. The same effect happens when an object is whirled around on the end of a string. The pull of the string on the object makes it go around in a circle.

Any object moving round a circular path is acted on by a resultant force which is directed towards the centre of the circle. The resultant force is called the **centripetal force** because it acts towards the centre of the circle.

- For a satellite in a circular orbit, the centripetal force is due to a single force, namely the force of gravity on the satellite.
- For an object being whirled around in a horizontal circle at the end of a string, the centripetal force is due to the pull of the string on the object. This force acts towards the hand holding the string at the centre of the circle.

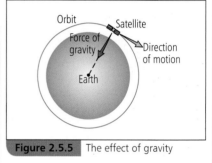

Figure 2.5.5 The effect of gravity

SUMMARY QUESTIONS

1 Copy and complete the following sentences using words from the list.

equal to   greater than   less than   not zero   zero

A car starts from rest and accelerates along a straight flat road until it reaches a certain speed which it then travels at.

a When it is accelerating, the force of its engine is _____ the resistive forces acting on it. The resultant force is _____ .

b When it is travelling at constant speed, the force of air resistance on it is _____ the force of its engine. The resultant force on the vehicle is _____ .

2 A stone of weight 1.5 N released at the surface of a swimming pool drops to the bottom of the pool at constant speed.

a What can you say about the resultant force on the stone as it descends?

b State the magnitude and direction of the resistive force on the stone as it descends.

3 a A lorry tows a car by means of a tow bar along a straight road at constant speed.

i What can you say about the acceleration of the lorry?

ii The force of the lorry on the car was 200 N. State the magnitude of resistive force on the car.

b The lorry driver applies the brakes to the lorry, causing the lorry and the car to slow down and stop. In terms of the forces on the car, explain why the car comes to a stop.

KEY POINTS

1 The resultant force due to two forces acting along the same line is given by:

- the sum of the two forces if the forces act in the same direction
- the difference between the two forces if they act in opposite directions.

2 When the resultant force is non-zero, the object experiences an acceleration given by:

$$\text{acceleration} = \frac{\text{resultant force}}{\text{mass}}$$

LEARNING OUTCOMES

- Recognise that momentum is defined as mass × velocity
- Recall that the unit of momentum is the kilogram metre per second (kg m/s)
- Use the equation 'momentum = mass × velocity'

A contact sport

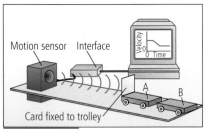

Figure 2.6.1 Investigating collisions

DID YOU KNOW?

If a vehicle crashes into the back of a line of cars, each car is shunted into the one in front. Momentum is transferred along the line of cars to the one at the front.

Momentum is important to anyone who plays a contact sport. In a game of rugby, a player with a lot of momentum is very difficult to stop.

The momentum of a moving object = its mass × its velocity.

So momentum has a magnitude and a direction.

The unit of momentum is the kilogram metre/second (kg m/s).

We can write the word equation using symbols: $p = m \times v$, where

p = momentum in kilogram metres/second, kg m/s

m = mass in kilograms, kg

v = speed in metres/second, m/s

PRACTICAL

Investigating collisions

When two objects collide, the momentum of each object changes. Figure 2.6.1 shows how to use a computer and a motion sensor to investigate a collision between two trolleys.

Trolley A is given a push so it collides with stationary trolley B. The two trolleys stick together after the collision. The computer gives the velocity of A before the collision and the velocity of both trolleys afterwards.

- What does each section of the velocity–time graph show?

1 **For two trolleys of the same mass,** the velocity of trolley A is halved by the impact. The combined mass after the collision is twice the moving mass before the collision. So the momentum (= mass × velocity) after the collision is the same as before the collision.

2 **For a single trolley pushed into a double trolley,** the velocity of A is reduced to one-third. The combined mass after the collision is three times the initial mass. So in this test as well, the momentum after the collision is the same as the momentum before the collision.

In both tests, the total momentum is unchanged (i.e. is conserved) by the collision. This is an example of the **conservation of momentum**. It applies to any system of objects as long as the system is closed, which means that no resultant force acts on it.

Safety: Use foam or an empty cardboard box to stop trolleys. Protect bench and feet from falling trolleys.

Figure 2.6.2 A 'shunt' collision

In general, the **law of conservation of momentum** states that:

In a closed system, the total momentum before an event is equal to the total momentum after the event.

We can use this law to predict what happens whenever objects collide or push each other apart in an 'explosion'. Momentum is conserved in any collision or explosion as long as no external forces act on the object.

WORKED EXAMPLE

A 0.5 kg trolley A is pushed at a velocity of 1.2 m/s into a stationary trolley B of mass 1.5 kg as shown in Figure 2.6.3. The two trolleys stick to each other after the impact.

a Calculate the momentum of the two 0.5 kg trolleys before the collision.

b Calculate the velocity of the two trolleys straight after the impact.

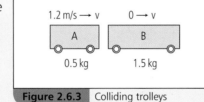

Solution

a Momentum = mass × velocity = 0.5 kg × 1.2 m/s = 0.6 kg m/s

Figure 2.6.3 Colliding trolleys

b The momentum after the impact = momentum before the impact = 0.6 kg m/s

(1.5 kg + 0.5 kg) × velocity after the impact = 0.6 m/s

The velocity after the impact = $\dfrac{0.6\,\text{kg m/s}}{2\,\text{kg}}$ = 0.3 m/s

WORKED EXAMPLE

A 3000 kg truck moving at a velocity of 16 m/s crashes into the back of a stationary 1000 kg car. The two vehicles move together immediately after the impact. Calculate their velocity.

Solution

Let v represent the velocity of the vehicles after the impact.

momentum of truck before impact = 48 000 kg m/s

momentum of the car before impact = 0 kg m/s

momentum of truck after impact = 3000 kg × v

momentum of car after impact = 1000 kg × v

$3000v + 1000v = 48\,000 + 0$

$4000v = 48\,000$; v = **12 m/s**

KEY POINTS

1 Momentum = mass × velocity

2 The unit of momentum is kg m/s.

3 Momentum is conserved whenever objects interact, as along as objects are in a closed system so that no external forces act on them.

SUMMARY QUESTIONS

1 a Define momentum and state its unit.

  b Calculate the momentum of a 40 kg person running at 6 m/s.

  c In the worked example shown by Figure 2.6.3, calculate the speed after the collision if trolley A had a mass of 1.0 kg.

2 a Calculate the momentum of an 80 kg rugby player running at a velocity of 5 m/s.

  b An 800 kg car moves with the same momentum as the rugby player in **a**. Calculate the velocity of the car.

  c Calculate the velocity of a 0.4 kg ball that has the same momentum as the rugby player in **a**.

3 A 1000 kg wagon moving at a velocity of 5.0 m/s on a level track collides with a stationary 1500 kg wagon. The two wagons move together after the collision.

  a Calculate the momentum of the 1000 kg wagon before the collision.

  b Calculate the velocity of the wagons after the collision.

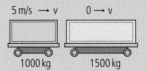

LEARNING OUTCOMES

- Recognise that momentum has magnitude as well as direction
- Recall that when two objects of different masses push each other apart, they move away at different speeds and have equal and opposite momentum

If you are skateboarder, you will know that the skateboard can shoot away from you when you jump off it. Its momentum is in the opposite direction to your own momentum. What can we say about the total momentum of objects when they fly apart from each other?

PRACTICAL

Investigating a controlled explosion

Figure 2.7.1 shows a controlled explosion using trolleys. When the trigger rod is trapped, a bolt springs out and the trolleys recoil (spring back) from each other.

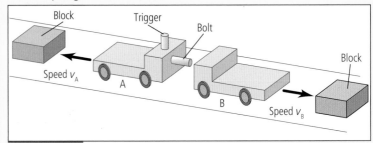

Figure 2.7.1 Investigating explosions

Using trial and error, we can place blocks on the runway so the trolleys reach them at the same time. This allows us to compare the speeds of the trolleys.

Some results are shown in Figure 2.7.2.

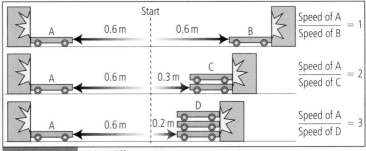

Figure 2.7.2 Using different masses

- Two single trolleys travel equal distances in the same time. This shows that they recoil at different speeds.
- A double trolley travels only half the distance that a single trolley does. Its speed is half that of the single trolley.

In each test:

1. The mass of the trolley × the speed of the trolley is the same.

2. They recoil in opposite directions.

So momentum has magnitude and direction. The results show that the trolleys recoil with equal and opposite momentum.

STUDY TIP

Be careful in calculations – momentum is a vector quantity, so if two objects are travelling in opposite directions, one has positive momentum, and the other has negative momentum.

Conservation of momentum

In the trolley examples:

- momentum of A after the explosion = (mass of A × velocity of A)
- momentum of B after the explosion = (mass of B × velocity of B)
- total momentum before the explosion = 0 (because both trolleys were at rest).

Using conservation of momentum gives:

(mass of A × velocity of A) + (mass of B × velocity of B) = 0

Therefore

(mass of A × velocity of A) = − (mass of B × velocity of B)

The minus sign after the equals sign tells us that the momentum of B is in the opposite direction to the momentum of A. The equation tells us that A and B move apart with equal and opposite amounts of momentum. So, the total momentum after the explosion is the same as before it.

Momentum in action

When a shell is fired from a military gun, the gun barrel recoils backwards. The recoil of the gun barrel is slowed down by a spring. This lessens the backwards motion of the gun.

WORKED EXAMPLE

An artillery gun of mass 2000 kg fires a shell of mass 20 kg at a velocity of 120 m/s. Calculate the recoil velocity of the gun.

Solution

Applying the conservation of momentum gives:

mass of gun × recoil velocity of gun = − (mass of shell × velocity of shell)

If we let V represent the recoil velocity of the gun,

$2000 \, \text{kg} \times V = -(20 \, \text{kg} \times 120 \, \text{m/s})$

$$V = -\frac{2400 \, \text{kg m/s}}{2000 \, \text{kg}} = -1.2 \, \text{m/s}$$

Figure 2.7.3 An artillery gun in action

SUMMARY QUESTIONS

1 A 60 kg skater and an 80 kg skater standing in the middle of an ice rink, push each other away.

  What can be said about:

  80 kg   60 kg

  a the force they exert on each other when they push apart?

  b the momentum each skater has just after they separate?

  c each of their velocities after they separate?

  d their total momentum just after they separate?

2 In Question 1, the 60 kg skater moves away at 2.0 m/s. Calculate:

  a her momentum

  b the velocity of the other skater.

3 A 600 kg cannon recoils at a speed of 0.5 m/s when a 12 kg cannonball is fired from it.

  a Calculate the velocity of the cannonball when it leaves the cannon.

  b How would the recoil velocity of the cannon have been different if a 4 kg cannonball had been used instead?

KEY POINTS

1 Momentum is mass × velocity and has direction.

2 When two objects push each other apart, they move:

  - with different speeds if they have unequal masses
  - with equal and opposite momentum, so their total momentum is zero.

LEARNING OUTCOMES

- Recognise that when objects collide, the force of the impact depends on how long the impact lasts for
- Recall that the impulse of a force F acting on an object for a time t is defined as $F \times t$
- Carry out calculations using the equation impulse = change of momentum

PRACTICAL

Investigating impacts

We can test an impact using a trolley and brick, as shown in Figure 2.8.1. When the trolley hits the brick, the plasticine flattens on impact, making the impact time longer. This is the key factor that reduces the impact force.

Safety: Protect bench and feet from bricks and trolley.

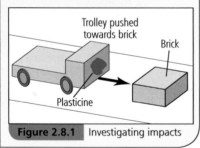

Trolley pushed towards brick

Brick

Plasticine

Figure 2.8.1 | Investigating impacts

STUDY TIP

Remember that the time of impact is important. Sometimes we want a force to be large – for example when hitting a ball – so the impact time should be as short as possible. At other times, we want the force to be small – for example in a crash – so the time of impact should be as long as possible.

Crumple zones at the front end and rear end of a car are designed to lessen the force of an impact. The force changes the momentum of the car.

- In a front-end impact, the momentum of the car is reduced.
- In a rear-end impact (where a vehicle is struck behind by another vehicle), the momentum of the car is increased.

In both cases the effect of a crumple zone is to increase the impact time and so lessen the impact force.

Impact time

Let's see why making the impact time longer reduces the impact force.

Suppose a moving trolley hits another object and stops. The impact force on the trolley acts for a certain time (the impact time) and causes it to stop. A soft pad on the front of the trolley would increase the impact time and would allow the trolley to travel further before it stops. The momentum of the trolley would be lost over a longer time and its kinetic energy would be transferred over a greater distance.

If we know the impact time t, we can calculate the impact force F as follows:

- From Topic 2.4, $F = ma$, where m is the mass of the object and a is its acceleration.
- From Topic 1.4, $a = \dfrac{v - u}{t}$, where u is the initial velocity, v is the final velocity and t is the impact time.

Therefore, $F = ma = m\dfrac{(v - u)}{t} = \dfrac{mv - mu}{t}$

So we can use the above equation to calculate F.

Also, multiplying both sides of the equation by t gives:

$$Ft = mv - mu$$

where the force $F \times$ the time t is called the **impulse** of the force. The equation above therefore tells us that:

The impulse of a force, Ft = the change of momentum ($mv - mu$).

The above equation shows that for a given change of momentum, the impact force can be reduced by increasing the impact time.

Figure 2.8.2 | A crash test. Car makers test the design of a crumple zone by driving a remote control car into a wall.

WORKED EXAMPLE

A bullet of mass 0.004 kg moving at a velocity of 90 m/s is stopped by a bulletproof vest in 0.0003 s. Calculate the impact force.

Solution

Initial momentum of bullet = mass × initial velocity
= 0.004 kg × 90 m/s = 0.36 kg m/s

Final momentum of bullet = 0 (as the bullet is stopped)

Therefore the change of momentum
= final momentum − initial momentum = −0.36 kg m/s

Since Ft = change of momentum, where F is the impact force and t is the time taken, then

$F \times 0.0003\,s = -0.36\,kg\,m/s$

Dividing both sides by 0.0003 s gives $F = \dfrac{-0.36\,kg\,m/s}{0.0003\,s} = -1200\,N$

The negative sign tells us that the impact force decelerates the bullet.

Two-vehicle collision

When two vehicles collide, they exert equal and opposite impact forces on each other at the same time. The change of momentum of one vehicle is therefore equal and opposite to the change of momentum of the other vehicle. The total momentum of the two vehicles is the same after impact as it was before the impact, so momentum is conserved – assuming no external forces act.

For example, suppose a fast-moving truck runs into the back of a stationary car. The impact decelerates the truck and accelerates the car. Assuming that the truck's mass is greater than the mass of the car, the truck loses momentum and the car gains momentum.

DID YOU KNOW?

Scientists at Oxford University have developed new lightweight material for bulletproof vests. The material is so strong and elastic that bullets bounce off it.

DID YOU KNOW?

We sometimes express the effect of an impact on an object or person as a force-to-weight ratio. We call this **g-force**.

You would experience a g-force of:

- about 3 to 4 g on a fairground ride that whirls you around
- about 10 g in a low-speed car crash
- more than 50 g in a high-speed car crash that you would be lucky to survive.

SUMMARY QUESTIONS

1 a In a car crash, when a passenger wears a seat belt, why does it reduce the impact force on him?

  b A ball of mass of 0.12 kg moving at a velocity of 18 m/s is caught by a person in 0.0003 s. Calculate the impact force.

2 a An 800 kg car travelling at 30 m/s is stopped safely when the brakes are applied. What braking force is required to stop it in 6.0 s?

  b If the vehicle in **a** had been stopped in a collision lasting less than a second, explain by referring to momentum why the force on it would have been much greater.

3 A 2000 kg van moving at a velocity of 12 m/s crashes into the back of a stationary truck of mass 10 000 kg. Immediately after the impact, the two vehicles move together.

  a Show that the velocity of the van and the truck immediately after the impact is 2 m/s.

  b The impact lasts for 0.3 seconds. Calculate:

  i the change of momentum of the van

  ii the force of the impact of the van.

KEY POINTS

- When vehicles collide, the force of impact depends on mass, change of velocity and the duration of the impact.
- The longer the impact time, the more the impact force is reduced.
- When two vehicles collide:
  - they exert equal and opposite forces on each other
  - their total momentum is unchanged.

1 The gravitational field strength at the Earth's surface is 10 N/kg. Calculate the weight of the following:

(a) a person of mass 80 kg

(b) a 2 kg bag of sugar

(c) a 125 g pack of tea

(d) a 70 g chocolate bar

2 (a) Write down the equation used for calculating density.

(b) Describe, with the aid of diagrams, the displacement method for finding the density of a small, irregularly shaped object.

3 A rectangular block has the dimensions shown in the diagram.

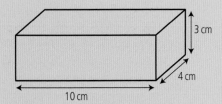

(a) Calculate the volume of the block.

(b) The block has a mass of 150 g. Calculate the density of the material of the block.

(c) Will this block float on water?

(d) Suggest a material from which the block might have been made.

4 (a) State the resultant force acting on a vehicle travelling at a constant speed of 120 km/h on a straight section of road.

(b)(i) A heavy truck is slowing down because the driver has applied the brakes. What can you conclude about the direction of the resultant force acting on the truck?

(ii) What is the resultant force acting on the truck when it has come to a halt?

5 (a) An object of mass 2.5 kg is acted on by a force of 5 N. Calculate the acceleration that this force produces.

(b) The same object is acted on by a different force and the acceleration is 3.2 m/s². Calculate the value of the force.

1 A student has an apple in his pocket. The weight of the apple is approximately

A 0.1 N

B 1 N

C 10 N

D 100 N

(Paper 1/2)

2 A car is travelling at a constant speed in a straight line on a motorway. The resultant force acting on the car is

A equal to its weight

B equal and opposite to the frictional force

C greater than the air resistance

D zero

(Paper 1/2)

3 A force of 120 N is applied to a mass of 4 kg. The acceleration, in m/s², of the mass is

A 0.033

B 30

C 480

D 4800

(Paper 2)

4 A toy car of mass 0.5 kg has a velocity of 4 m/s. Its momentum in SI units is

A 0.125

B 2

C 4

D 8

(Paper 2)

6 A stone is tied firmly to a length of string. The stone is then whirled around in a horizontal circle. The speed at which the stone is moving is constant.

(a) State whether or not the velocity of the stone is constant. Briefly explain your answer.

(b) The stone is accelerating. Explain how it can be accelerating whilst moving at constant speed.

(c) The acceleration is caused by a force. State the direction of the force and what type of force it is.

Supplement

5 The diagram shows the apparatus used for an experiment in which a steel spring is stretched.

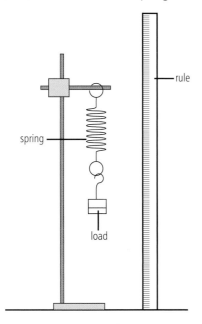

(a) The original length of the spring, before a load is hung on it, is 17 mm. When a load of 1.0 N is hung on the spring, the length is 20 mm.

(i) Calculate the extension of the spring in mm.

(ii) The load is increased to 2.0 N and the length increases to 23 mm. Calculate the extension of the spring in mm. *[2]*

(b) The load is increased to 4.0 N. The spring does not overstretch. Calculate the extension with the 4.0 N load. *[2]*

(c) Name the instrument that uses a spring to measure weight. *[1]*

(Paper 3)

6 The speed–time graph shows the motion of a car travelling in a straight line.

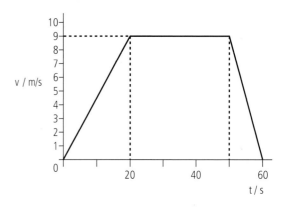

(a) Describe the motion of the car

(i) during the first 20 s

(ii) during the next 30 s

(iii) during the final 10 s. *[4]*

(b) Calculate the acceleration of the car during the first 20 s. *[2]*

(c) The car has a mass of 800 kg. Calculate the resultant force acting on the car during the first 20 s. *[2]*

(d) Calculate the distance travelled by the car in 60 s. *[2]*

(Paper 3)

7 A car is travelling at 8 m/s. The mass of the car is 750 kg.

(a) Calculate the momentum of the car. *[2]*

(b) Calculate the force required to stop the car in 8 s. *[2]*

(c) Calculate the force required to stop the car in 2 s. *[1]*

(Paper 4)

3 Forces in equilibrium

3.1 Moments

LEARNING OUTCOMES

- Describe what is meant by the moment of a force about a point
- Recognise and describe everyday examples of moments
- Describe the balancing of a beam about a pivot

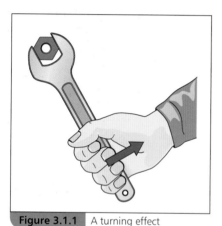

Figure 3.1.1 A turning effect

DID YOU KNOW?

A patient fitted with a replacement hip joint has to be very careful at first. A slight movement can cause a turning effect that pulls the hip joint apart.

Using moments

A **spanner** is needed to undo a very tight wheel nut on a bicycle. The force you apply to the spanner has a turning effect on the nut. You couldn't undo a tight nut with your fingers but the spanner can turn it. The spanner exerts a much bigger force on the nut than the force you apply to the spanner.

If you had a choice between a long-handled spanner and a short-handled one, which would you choose? The longer the spanner's handle, the less force you need to exert on it to untighten the nut (Figure 3.1.1). In this example, the turning effect of the force, called the **moment** of the force, can be increased by:

- increasing the size of the force
- using a spanner with a longer handle.

Investigating the turning effect of a force

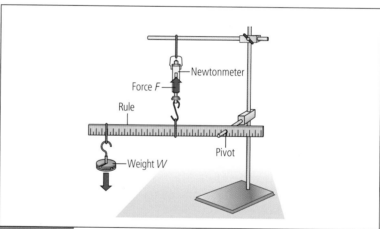

Figure 3.1.2 Investigating turning forces

Figure 3.1.2 shows one way to investigate the turning effect of a force. A known weight W is suspended from the metre rule as shown and then moved along it until the rule is exactly horizontal.

- How do you think the reading on the newtonmeter compares with the weight W? You should find that the reading (i.e. the force needed to support the weight) is greater than W, provided W is further from the pivot than the newtonmeter is. The rule is in equilibrium because the turning effect of the weight is equal and opposite to the turning effect of the newtonmeter.
- If the weight is moved further from the pivot, the free end of the rule drops down. This shows that the turning effect of the weight increases the further the weight is from the pivot.

The **moment of a force** is given by the equation:

moment	= force	×	perpendicular distance to pivot
(in newton metres, N m)	(in newtons, N)		(in metres, m)

Note that the unit of the moment of a force is the newton metre (N m).

The **claw hammer** in Figure 3.1.3 is being used to remove a nail from a wooden beam.

- The applied force F on the claw hammer tries to turn it clockwise about the pivot.
- The moment of force F about the pivot is F × d, where d is the perpendicular distance from the pivot to the line of action of the force.
- The effect of the moment is to cause a much larger force to be exerted on the nail.

STUDY TIP

The moment of a force is an important concept – the next time you close a car door, remember that less force is needed if you push on the door as far from the hinges as possible.

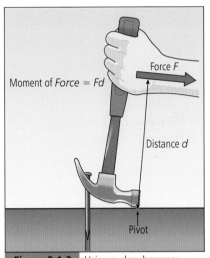

Moment of *Force* = *Fd*

Force *F*

Distance *d*

Pivot

Figure 3.1.3 Using a claw hammer

WORKED EXAMPLE

A force of 50 N is exerted on a claw hammer of length 0.30 m, as shown in Figure 3.1.3. Calculate the moment of the force.

Solution

Force = 50 N × 0.30 m = 15 N m

SUMMARY QUESTIONS

1 In Figure 3.1.1, a force is applied to a spanner to undo a nut. State whether the moment of the force is:

  a clockwise or anticlockwise

  b increased or decreased by **i** increasing the force, **ii** exerting the force nearer the nut.

2 Explain each of the following statements:

  a A claw hammer is easier to use to remove a nail if the hammer has a long handle.

  b A door with rusty hinges is more difficult to open than a door of the same size with lubricated hinges.

3 A spanner is used to tighten a nut as shown in Figure 3.1.1. A force of 50 N is exerted on the spanner at a distance of 0.24 m from the centre of the nut. Calculate the moment of the force.

KEY POINTS

The moment of a force F about a pivot is F × d, where d is the perpendicular distance from the pivot to the line of action of the force.

LEARNING OUTCOMES

- Recognise that there is no resultant force or resultant turning effect on an object in equilibrium
- Use knowledge of forces and turning effects to explain why objects at rest don't move or turn

STUDY TIP

Remember the resultant force on an object is the effect of all the forces acting on it.

KEY POINTS

There is no resultant force or resultant turning effect on an object in equilibrium.

On a seesaw

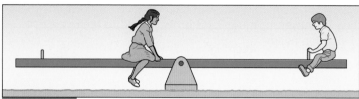

Figure 3.2.1 Moments in action – the seesaw

A seesaw is an example in which clockwise and anticlockwise moments balance each other out. The girl in Figure 3.2.1 sits near the pivot to balance her younger brother at the far end of the seesaw. He is not as heavy as his big sister.

- His weight has a clockwise turning effect because it would make the seesaw turn clockwise if the girl were not on the seesaw.
- Her weight has an anticlockwise turning effect because it would make the seesaw turn anticlockwise if the boy were not on the seesaw.

Because the turning effect of each child depends on the child's weight and their distance to the pivot, the girl needs to sit nearer the pivot than her brother so that their turning effects about the pivot balance out. Therefore there is no resultant turning effect on the seesaw.

The pivot supports the weight of the two children and the weight of the seesaw beam. The support force acting upwards on the beam from the pivot must be equal and opposite to the total weight of the beam and the children. Therefore there is no resultant force on the seesaw beam.

On a building site

Figure 3.2.2 shows a horizontal plank supported near each end by steel scaffolding tubes X and Y. A builder stands on the plank between X and Y.

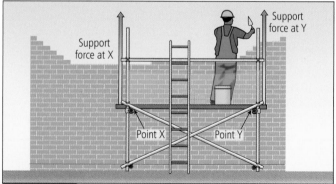

Figure 3.2.2 Moments at work – on a building site

- The tubes X and Y support the total weight of the plank and the builder. The support forces at X and Y:
  - act vertically upwards from the tubes X and Y
  - are equal to the total weight of the builder and the plank (which act downwards).

The resultant force on the plank is therefore zero.

- The support force at X is equal to the support force at Y if the builder is midway between X and Y, and the middle of the plank is also midway between X and Y.

The resultant turning effect on the plank is therefore zero.

These two examples show that, for any object in equilibrium:

- **there is no resultant turning effect on it**
- **there is no resultant force on it.**

PRACTICAL

Investigating a toy mobile

1 Measure the weight of a rule and of two small objects A and B, using a newtonmeter. Make a model of a toy mobile using the rule and the two objects as shown in Figure 3.2.3 and suspend it from a newtonmeter. Note that the point of suspension S of the rule is at its centre.

2 Adjust the positions of the two objects A and B so the rule is horizontal.

- You should find that the heavier of the two objects is nearer S than the other object. This is because the two objects have equal and opposite turning effects on the rule when the rule is horizontal. So the resultant turning effect on the mobile is zero.

- You should also find the newtonmeter reading is equal to the sum of the weight of the rule and the weights of the two objects. This is because the support force from the newtonmeter on the rule acts vertically upwards and is equal and opposite to the total weight of the mobile. So there is no resultant force on the mobile.

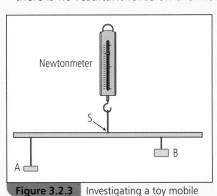

Figure 3.2.3 Investigating a toy mobile

1 a In the seesaw shown in Figure 3.2.1, state and explain whether the seesaw will turn clockwise or anticlockwise if the girl shifts away from the pivot.

b Figure 3.2.4 below shows a simple model of a seesaw in which a rule pivoted at its centre supports two weights A and B. The forces acting on the rule are shown in Figure 3.2.4.

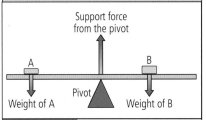

Figure 3.2.4

i When the rule is in equilibrium, A is further from its centre than B. What does this tell you about the weight of A compared with the weight of B?

ii What can you say about the support force on the rule from the pivot compared with the weights of A and B?

2 In Figure 3.2.2, the total weight of the plank and the builder is 860 N.

a The builder is at the centre of the plank which is midway between the supports X and Y. Calculate the support force from each support.

b State and explain how each support force changes when the builder moves nearer to Y.

Supplement

LEARNING OUTCOMES

- Describe and calculate the moment of a force
- Apply the principle of moments to a beam balanced about a pivot
- Apply the principle of moments to different situations

STUDY TIP

There are many examples of turning effects – far more than can be given here. Use the examples here to make sure you understand the principles so you can apply them to other examples.

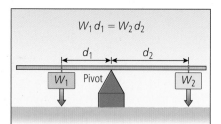

Figure 3.3.2 The Principle of Moments

Turning effects at work

The next time you have to move a heavy load, think beforehand about how to make the job easier. Use the examples here to make sure you understand the principles so you can apply them to other examples. The load (weight W_0) is lifted and moved using a much smaller effort (force F_1). This is because:

- the pivot is the point where the wheel is in contact with the ground
- the turning effect of the effort about the pivot is equal and opposite to the turning effect of the load about the pivot
- the effort acts much further from the pivot than the load so the effort needed to raise the load is much smaller than the load.

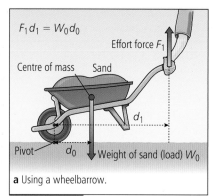
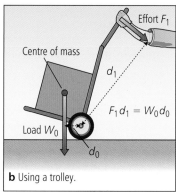

a Using a wheelbarrow.

b Using a trolley.

Figure 3.3.1 Moments at work

Balancing a beam

Figure 3.3.2 shows two weights W_1 and W_2 on a pivoted metre rule in equilibrium. Before placing the weights on the rule, the rule on its own was balanced horizontally on the pivot. With W_1 at a certain distance d_1 from the pivot, the distance d_2 of W_2 from the pivot is adjusted until the rule is horizontal. The turning effect of W_1 about the pivot is then equal and opposite to the turning effect of W_2 about the pivot.

The Principle of Moments

Measurements obtained using the arrangement in Figure 3.3.2 with different weights and distances are shown in the table below. Each row of measurements shows that, in each case, the moment of W_1 about the pivot ($= W_1 \times d_1$) is equal and opposite to the moment due to W_2 ($= W_2 \times d_2$).

weight W_1 / N	weight W_2 / N	distance d_1 / m	distance d_2 / m	$W_1 \times d_1$ / N m	$W_2 \times d_2$ / N m
1.0	2.0	0.48	0.24	0.48	0.48
1.0	3.0	0.48	0.16	0.48	0.48
1.0	4.0	0.48	0.12	0.48	0.48

The examples in the table demonstrate the **Principle of Moments**. This states that, **for an object in equilibrium**:

the sum of all the clockwise moments about any point = the sum of the all anticlockwise moments about that point.

PRACTICAL

Moments tests

1 **Measuring an unknown weight**

We can use the arrangement in Figure 3.3.2 to find an unknown weight, W_1, if we know the other weight, W_2, and we measure the distances d_1 and d_2. The unknown weight can then be calculated using the equation $W_1d_1 = W_2d_2$.

2 **Three weights on a beam**

Check the rule on its own is balanced on the pivot. Place known weights W_1, W_2 and W_3 on the beam, as in Figure 3.3.3, with W_2 between W_1 and the pivot, and the pivot at the same position as before. Rebalance the beam by adjusting the positions of the weights. Measure the distances d_1, d_2 and d_3 from each weight to the pivot.

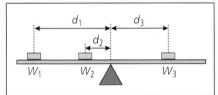

Figure 3.3.3

• The total anticlockwise moment about the pivot = the moment of W_1 + the moment of $W_2 = W_1d_1 + W_2d_2$.
• The total clockwise moment about the pivot = the moment of $W_3 = W_3d_3$.

Applying the Principle of Moments gives $W_1d_1 + W_2d_2 = W_3d_3$.

Use your measurements to confirm the above equation.

WORKED EXAMPLE

Calculate W_1 in Figure 3.3.3 if $W_2 = 2.0\,N$, $W_3 = 4.0\,N$, $d_1 = 0.25\,m$, $d_2 = 0.10\,m$ and $d_3 = 0.20\,m$.

Solution

Using $W_1d_1 + W_2d_2 = W_3d_3$ gives:
$(W_1 \times 0.25) + (2.0 \times 0.10) = (4.0 \times 0.20)$

Therefore $0.25W_1 + 0.20 = 0.80$

$0.25W_1 = 0.80 - 0.20 = 0.60$

$W_1 = \dfrac{0.60}{0.25} = 2.4\,N$

KEY POINTS

For an object in equilibrium:

the sum of the anticlockwise moments about any point = the sum of the clockwise moments about that point.

SUMMARY QUESTIONS

1 Dawn sits on a seesaw 2.50 m from the pivot. Jasmin balances the seesaw by sitting 2.00 m on the other side of the pivot.

 a Who is lighter, Dawn or Jasmin?

 b Dawn picks up her younger brother who weighs less than she does and sits him on the seesaw with her. Explain why Jasmin needs to move further away from the pivot to rebalance the seesaw.

2 a For the balanced beam in Figure 3.3.4, work out the unknown weight, W.

Figure 3.3.4

 b Figure 3.3.5 shows three weights on a beam that is balanced at its centre. Calculate the distance d from the 0.5 N weight to the pivot.

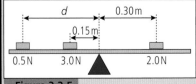

Figure 3.3.5

LEARNING OUTCOMES

- Describe what is meant by the centre of mass of a body
- Recognise that the centre of mass of a freely suspended object at rest is directly below the point of suspension
- Locate by experiment the centre of mass of a flat object

STUDY TIP

Make sure you know what is meant by 'equilibrium position'.

DID YOU KNOW?

A tightrope walker carries a long pole to keep balanced. The pole is used to keep the centre of mass of the walker (and pole) directly above the rope. A slight body movement one way is counterbalanced by shifting the pole slightly the other way.

The design of racing cars has changed considerably since the first models, but one design feature that has not changed is the need to keep the car near the ground. The weight of the car must be as low as possible, otherwise the car will overturn when cornering at high speeds.

Modern racing car design

We can think of the weight of an object as if it acts at a single point. This point is called the **centre of mass** (or the centre of gravity) of the object.

Every object behaves as if its mass were concentrated at one point. This point is called the centre of mass.

Suspended equilibrium

If you suspend an object and then release it, it will come to rest with its centre of mass directly below the point of suspension, as shown in Figure 3.4.1a. The object is then in **equilibrium.** Its weight does not exert a turning effect on the object because its centre of mass is directly below the point of suspension.

If the object is turned from this position and then released, it will swing back to its equilibrium position. This is because its weight has a turning effect that returns the object to equilibrium, as shown in Figure 3.4.1b.

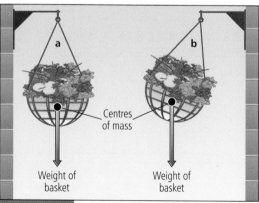

Centres of mass

Weight of basket

Weight of basket

Figure 3.4.1 Suspension (a) in equilibrium, (b) non-equilibrium

Locating the centre of mass of a flat object

Figure 3.4.2 shows how to find the centre of mass of a flat card.

1 Suspend the card freely from a thin rod and allow it to come to rest. Its centre of mass is then directly below the rod. Use a 'plumbline' to draw a vertical line on the card from the thin rod downwards. To do this, mark the bottom edge of card where the plumbline meets the edge. Remove the plumbline and card from the pivot. Draw a line joining the mark to the pivot hole.

2 Repeat the procedure with the card suspended from a second point to give another similar line. The centre of mass of the card is where the two lines meet.

3 Test your result by suspending the card from a third point and using the plumbline to draw a vertical line through this third point. The third line should pass through the point where the other two lines meet.

See if you can balance the card at this point on the end of a pencil.

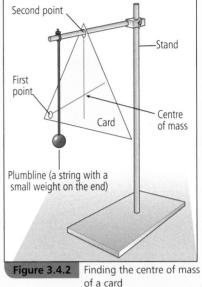

Figure 3.4.2 Finding the centre of mass of a card

The centre of mass of a symmetrical object

For a symmetrical object, its centre of mass is along the axis of symmetry, as shown in Figure 3.4.3. If the object has more than one axis of symmetry, its centre of mass is where the axes of symmetry meet.

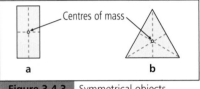

Figure 3.4.3 Symmetrical objects

- A rectangle has two axes of symmetry, as shown in Figure 3.4.3a. The centre of mass is where the axes meet.
- The equilateral triangle in Figure 3.4.3b has three axes of symmetry, each bisecting one of the angles of the triangle. The three axes meet at the same point, which is the centre of mass of the triangle.

STUDY TIP

Make sure you can describe all the steps in the above experiment.

1 a On a tightrope, a tightrope walker carrying a pole horizontally senses a slight movement to the left. Should the pole be shifted to the left or right? Give a reason for your answer.

b Explain why the centre of mass of a spoon is not at its middle.

2 Sketch each of the objects shown in Figure 3.4.4 and mark its centre of mass.

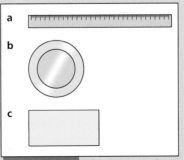

Figure 3.4.4 (a) A 300 mm rule, (b) a circular plate, (c) a rectangle

KEY POINTS

1 Every object behaves as if its mass were concentrated at one point. This point is called its centre of mass.

2 When a suspended object is in equilibrium, its centre of mass is directly beneath the point of suspension.

3 The centre of mass of a symmetrical object is along the axis of symmetry.

Stability

LEARNING OUTCOMES

- Recognise the factors that affect the stability of an object
- Explain why an unstable body may topple over

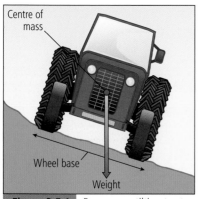

Figure 3.5.1 Forces on a tilting tractor

Centre of mass

Wheel base

Weight

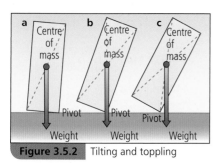

Figure 3.5.2 Tilting and toppling

a Centre of mass Pivot Weight

b Centre of mass Pivot Weight

c Centre of mass Pivot Weight

A toppling test

Stability and safety

Look around you and see how many objects could topple over. Bottles, table lamps and floor-standing bookcases are just a few objects that can easily topple over. Lots of objects are designed for stability so they can't topple over easily.

1 Tractor safety

Figure 3.5.1 shows a tractor on a hillside. It doesn't topple over because the line of action of its weight acts within its wheelbase. If it is tilted more, it would topple over if the line of action of its weight acted outside its wheelbase. Its weight would then give a clockwise turning effect about the lower wheel.

2 Bus tests

The photo shows a double-decker bus being tested to see how much it can tilt without toppling over. Such tests are important to make sure buses are safe to travel on, especially when they go around bends and on hilly roads.

3 Ladders

If you climb a ladder, don't lean too far from it. If you do, you might shift your centre of mass too far and topple over.

PRACTICAL

Tilting and toppling tests

How far can you tilt something before it topples over? Figure 3.5.2 shows how you can test your ideas using a tall box or a brick on its end.

1 If you tilt the brick slightly, as in **a**, and release it, the turning effect of its weight returns it to its upright position.

2 If you tilt the brick more, you can just about balance it on one edge, as in **b**. Its centre of mass is then directly above the edge on which it balances. Its weight has no turning effect in this position.

3 If you tilt the brick even more, as in **c**, it will topple over if it is released. This is because the line of action of its weight is 'outside' its base. So its weight has a turning effect that makes it topple over.

PRACTICAL

Measuring the weight of a beam

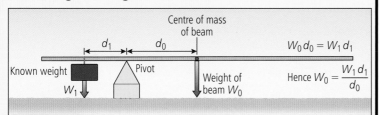

Figure 3.5.3 Finding the weight of a beam

1 A beam placed off-centre on a pivot will topple off the pivot because it is unstable.

2 Figure 3.5.3 shows how to balance a beam off-centre on a pivot using an object as a 'counterweight'. The weight of the beam acts at its centre of mass which is at distance d_0 from the pivot.

- The moment of the beam about the pivot = $W_0 d_0$ clockwise, where W_0 is the weight of the beam.
- The moment of W_1 about the pivot = $W_1 d_1$ anticlockwise, where d_1 is the perpendicular distance from the pivot to the line of action of W_1.

Applying the principle of moments gives $W_1 d_1 = W_0 d_0$.

So we can calculate the beam's weight W_0 if we know W_1 and distances d_1 and d_0.

For example, if the weight $W_1 = 2.0\,N$, $d_1 = 0.15\,m$ and $d_0 = 0.25\,m$, prove for yourself that $W_0 = 1.2\,N$.

SUMMARY QUESTIONS

1 a Make a list of objects that are designed to be difficult to knock over.

b A well-designed laboratory stool has a base that is wider than the seat. If the base was too narrow, why would the seat be unsafe?

c Would a double-decker bus be more or less stable if everyone on it sat on the top deck?

2 The following measurements were made as shown in Figure 3.5.3 to find the weight of a beam.

$W_1 = 2.4\,N$, $d_1 = 0.10\,m$, $d_0 = 0.30\,m$.

Use these measurements to calculate:

a the weight of the beam

b the support force on the beam from the pivot.

STUDY TIP

Make sure you understand the idea of stability so that you can explain even unfamiliar examples.

DID YOU KNOW?

When you carry a heavy rucksack, don't sit on a wall. If you do, and you lean back, you'll topple over.

KEY POINTS

1 The stability of an object is increased by making its base as wide as possible and its centre of mass as low as possible.

2 An object will tend to topple over if the line of action of its weight is outside its base.

Supplement

LEARNING OUTCOMES

- Use the parallelogram of forces rule to find the resultant of two forces that do not act along the same line

- Describe an experiment to test the parallelogram of forces

- Recognise the difference between a vector and a scalar quantity

Force

We saw in Topic 2.5 that when two forces act along the same line on an object, the resultant force (i.e. their combined effect) is equal to:

- the sum of the two forces if the forces are in the same direction
- the difference between the two forces if the forces are in opposite directions.

What if the two forces do not act along the same line, as shown in the photograph opposite? This shows a ship being towed by cables from two tugboats. The tension force in each cable pulls on the ship. The combined effect of these tension forces is to pull the vessel as in Figure 3.6.1. This is the resultant force.

- Figure 3.6.1 shows how the two tension forces T_1 and T_2 in the tow ropes can be represented as vectors and combined to produce the resultant force. The tension forces are drawn as adjacent sides of a parallelogram; the resultant force is the diagonal of the parallelogram from the origin of T_1 and T_2. This geometrical method is called the **parallelogram of forces**.

Under tow

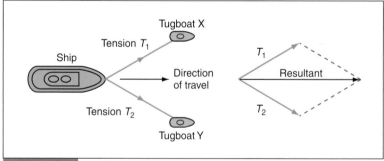

Figure 3.6.1   Combining forces

PRACTICAL

Investigating the parallelogram of forces

We can use weights and pulleys to demonstrate the parallelogram of forces, as shown in Figure 3.6.2. The tension in each string is equal to the weight it supports, either directly or over a pulley.

The point where the three strings meet is in equilibrium. The string supporting the middle weight (W_3) is vertical. The angles θ_1 and θ_2 between each of the other two strings and the vertical string are measured using a protractor. Using these measured angles, and the values of the three known weights, a scale diagram of a parallelogram is drawn, such that:

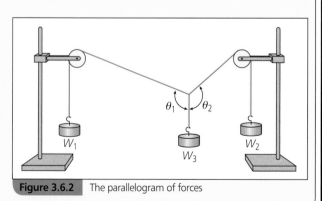

Figure 3.6.2   The parallelogram of forces

- the line down the centre of the diagram represents a vertical line
- adjacent sides of the parallelogram at angles θ_1 and θ_2 to the 'vertical' line represent the tensions in the strings supporting W_1 and W_2.
- The resultant of W_1 and W_2, represented by the diagonal, should be equal and opposite in direction to the vector representing W_3.

KEY POINTS

1 The parallelogram of forces is used to find the resultant of two forces that do not act along the same line.

2 A vector is a physical quantity that has a direction.

3 A scalar is a physical quantity that does not have direction.

WORKED EXAMPLE

A tow rope is attached to a car at two points 0.8 m apart. The two sections of rope joined to the car are the same length and are at 30° to each other, as shown in Figure 3.6.3. The pull on each attachment should not exceed 3000 N. Use the parallelogram of forces to determine the maximum tension in the main tow rope.

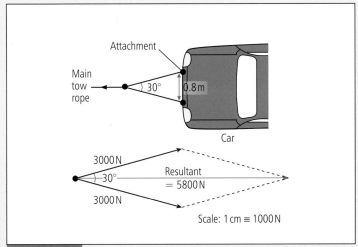

Figure 3.6.3 Using the parallelogram rule

Solution

The maximum tension T in the main tow rope is the resultant of the two 3000 N forces at 30° to each other. Drawing the parallelogram of forces as shown in Figure 3.6.3 gives $T = 5800\,\text{N}$.

SUMMARY QUESTIONS

1 Figure 3.6.4 shows two forces acting on an object X. Work out the magnitude and direction of the resultant force on X.

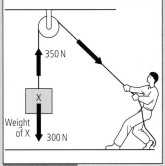

Figure 3.6.4

2 A force of 3.0 N and a force of 4.0 N act on a point object. Determine the magnitude of the resultant of these two forces if the angle between their lines of action is:

   a 90°, b 60°, c 45°.

Vectors and scalars

Many physical quantities in addition to force are directional. Physical quantities that are directional are called **vectors**. Examples include velocity, acceleration, weight and gravitational field strength.

Physical quantities that are not directional are called **scalars.** Examples include speed, mass, energy and power (which we will meet in the next chapter).

STUDY TIP

The distinction between vectors and scalars is important.

SUMMARY QUESTIONS

1 (a) Write the equation for the moment of a force.

(b) The diagram shows a force F being exerted by pulling on a rope to open a trap door. The weight of the trap door is 120 N. The trap door is square and the sides are of length 0.7 m. The centre of mass is at the centre of the square.

(i) Calculate the value of the force. Show your working.

(ii) What would be the value of the force if the rope were attached in the middle of the trap door instead of at the end?

2 A builder has two pairs of pliers. One has longer handles than the other. The builder applies the same force on each pair.

Explain which pair of pliers will exert a bigger force on the object in the jaw. Use the Principle of Moments in your explanation. It will probably help to include a labelled diagram in your answer.

3 (a) Suggest two household items that are designed to be stable.

(b) With the aid of suitable, labelled diagrams explain the features of the items that you have chosen that make them stable.

Supplement

4 (a) Speed is a scalar quantity and velocity is a vector quantity. Explain briefly the terms scalar and vector.

(b) Write down two more examples of scalar quantities and two more examples of vector quantities.

(c) A large object is to be pulled along using two ropes that are attached at the same point. The angle between the two ropes is 30°. The force exerted in one rope is 500 N and in the other is 600 N. Calculate the resultant force acting on the object. Show your working.

PRACTICE QUESTIONS

Q1, 2, 4, 5 (Paper 1/2); Q3 (Paper 2)

1 The unit of the moment of a force is

A N　**B** Nm　**C** N/m　**D** N/m²

2 The diagram shows a metre rule balanced at its midpoint on a pivot.

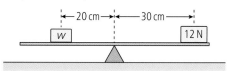

The value of the weight, w, is

A 8 N　**B** 12 N　**C** 18 N　**D** 22 N

3 Which of the following is NOT a vector?

A Acceleration　　**C** Velocity

B Mass　　**D** Weight

4 To achieve maximum stability, an object should have which combination of features?

A A high centre of mass and a narrow base

B A high centre of mass and a wide base

C A low centre of mass and a narrow base

D A low centre of mass and a wide base

5 Which of the following describes a large moment?

A A large force at a long distance from a pivot

B A large force at a short distance from a pivot

C A small force at a long distance from a pivot

D A small force at a small distance from a pivot

6 The diagram shows a beam pivoted at its centre of mass. Three equal loads A, B and C are hanging from the beam.

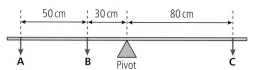

(a) (i) Which load, A, B or C, is exerting the smallest moment about the pivot?

(ii) Which load(s) is (are) exerting a clockwise moment? *[2]*

(b) Which one of these is a correct statement?

(i) The beam is balanced.

(ii) The beam tips so that the right hand end goes down.

(iii) The beam tips so that the right hand end goes up. *[1]*

(c) The diagram shows a wooden wedge. Copy the diagram and mark with a cross (×) the approximate position of the centre of mass. *[2]*

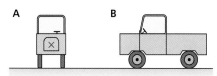

(Paper 3)

7 The diagrams A and B show the front and side views of a truck.

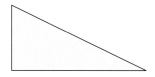

On diagram A the point X shows the height of the centre of mass above the ground. The driver and a passenger now get into the front seats.

(a) (i) Copy diagram A and mark with the letter Y a possible new position of the centre of mass.

(ii) Will the stability of the truck now be less, more or the same? *[3]*

(b) The truck is now loaded with some bricks, some plastic pipes and some insulating foam.

(i) State where you would position each of these in the back of the truck to keep it as stable as possible.

(ii) Explain your answer with reference to the position of the centre of mass. *[4]*

(c) Diagram C shows the truck on a slope. The point X shows the height of the centre of mass above the ground.

Explain why it is about to topple over. *[2]*

(Paper 4)

8 (a) State the Principle of Moments.

The diagram shows a uniform metre rule placed on a pivot. The pivot is at the centre of the rule. Two loads are placed on the metre rule in the positions shown.

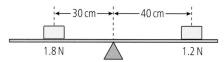

(b) (i) Calculate the clockwise and anticlockwise moments about the pivot. Show your working.

(ii) Explain why the rule is not balanced. *[3]*

(c) A third load of 0.6 N is available to balance the rule. (The positions of the other two loads and the pivot are not changed.) Determine where the 0.6 N load must be placed to balance the rule. Show your working. *[3]*

(Paper 3)

9 A student is provided with a boss, stand, clamp, plumbline, pin and a sheet of card. The shape of the sheet of card is shown in the diagram.

(a) (i) Draw a diagram to show how you would arrange the apparatus for an experiment to find the centre of mass of the card.

(ii) Describe briefly how you would carry out the experiment. *[5]*

(b) The diagram shows the plumbline. Two forces act on the metal sphere.

(i) Copy the diagram and label each of the arrows with the name of the force that it represents.

(ii) Copy and complete this sentence:

The two forces are _____ in size and _____ in direction.

(iii) Forces are vectors. State the meaning of the term 'vector'. *[6]*

(Paper 4)

4 Energy

4.1 Energy transfers

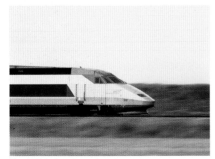

The TGV electric train

STUDY TIP

You must learn the names and descriptions of the forms of energy.

DID YOU KNOW?

The next time you ride a bike, think about what happens to your energy as you ride along. Most of it can never be used usefully again. But make sure you watch where you are going!

On the move

Cars, buses, planes and ships all use energy from fuel. They carry their own fuel. Electric trains use energy from fuel in power stations. Electricity transfers energy from the power station to the train.

We describe energy stored or transferred in different ways. The table below shows some of the different ways in which energy can be stored or transferred.

form of energy	description	example
kinetic energy	energy of an object due to its motion	vehicle in motion has kinetic energy
gravitational potential energy	energy of an object due to its position	book lifted up gains gravitational potential energy
chemical energy	energy stored in a substance and released when chemical reactions take place	car battery
electrical energy	energy transferred by an electric current	electric heater switched on
elastic strain energy	energy stored in an elastic object when we stretch or squash it	stretched spring
nuclear energy	energy released when the nucleus of an atom splits or disintegrates	A uranium fuel rod in a nuclear reactor
internal energy	energy of an object due to the internal motion and positions of its molecules	magnetised object or a hot object (Note: the energy of an object due to its temperature is sometimes referred to as *thermal energy*)
heat energy	energy transfer from a hot object to a cold object	heat radiation from burning coals
sound energy	energy transfer by sound waves	sound from a drum
light energy	energy transfer by light	light from a torch

Examples of energy transfers

Some energy transformations are described on the next page.

An athlete performing a pole vault uses his or her muscles to convert chemical energy into kinetic energy and elastic strain energy of the pole. This changes into gravitational potential energy, and also into heat energy and sound energy. The energy changes are:

chemical energy → kinetic energy + elastic energy →
gravitational potential energy (+ heat energy + sound energy)

A pile-driver on a building site is used to make firm foundations for tall buildings. Engineers use a pile-driver to hammer steel girders end-on into the ground. The pile-driver lifts a heavy steel block above the top end of the girder then lets it crash down onto the girder. The engineers keep doing this until the bottom end of the girder reaches solid rock. The energy changes of the block from its release to when it hits the girder are:

gravitational potential energy → kinetic energy → heat energy and
sound energy on impact

A pole vaulter

A pile-driver in action

PRACTICAL

Energy changes

1 Drop a small object onto a pad or cushion on the floor and observe the energy changes. As the object falls, it gains kinetic energy because it speeds up as it falls. So its gravitational potential energy changes to kinetic energy as it falls. On impact, it stops so what happens to the kinetic energy it had just before it hit the pad?

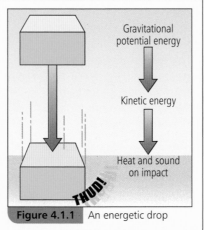

Figure 4.1.1 An energetic drop

2 Figure 4.1.1 shows a box that hits the floor with a thud. All its kinetic energy changes to heat and sound in the impact. Where does all the heat energy and sound energy go?

STUDY TIP

The amount of sound energy is very small compared to the amount of heat energy in these examples.

SUMMARY QUESTIONS

1 Choose words from the list below for the spaces in the sentences.

electrical   kinetic   gravitational potential   thermal

a When a ball falls in air, it loses _____ energy and gains _____ energy.

b When an electric heater is switched on, it changes _____ energy into _____ energy.

2 a List two different objects you could use to light a room in the event of a power cut. For each object, describe the energy changes that happen when it produces light.

b Which of the two objects in **a** is:
i easier to obtain energy from? ii easier to use?

KEY POINTS

1 Energy can be stored in objects in different ways and transferred in different ways.

2 Energy transferred to an object can be stored in different ways to the ways in which it was stored before being transferred.

Supplement

LEARNING OUTCOMES

- Describe the energy changes of an object that goes up and down
- Describe what is meant by conservation of energy and apply it to simple examples
- Explain why conservation of energy is a very important idea
- Apply conservation of energy to examples with multiple stages

On a roller coaster

STUDY TIP

Conservation of energy is one of the most fundamental principles in Physics!

At the fairground

Fairgrounds are very exciting places because lots of energy changes happen quickly. A roller coaster gains gravitational potential energy when it ascends and loses it when it descends.

As it descends the energy changes are:

gravitational potential energy → kinetic energy + sound + thermal energy due to air resistance and friction

PRACTICAL

Investigating energy changes

When energy changes happen, does the total amount of energy stay the same? We can investigate this question with a simple pendulum. Figure 4.2.1 shows a pendulum bob swinging from one side to the other.

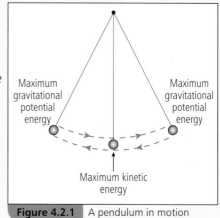

- As it moves towards the middle, its gravitational potential energy changes to kinetic energy.

Figure 4.2.1   A pendulum in motion

- As it moves away from the middle, its kinetic energy changes back to gravitational potential energy.

The principle of conservation of energy

Scientists have done lots of tests to find out if the total energy after a change is the same as before. All the tests so far show that it is the same.

The total amount of energy before and after a change is the same.

This important result is known as **conservation of energy**. It means that energy cannot be created or destroyed.

But energy tends to spread out, for example where energy is transferred to the surroundings by heating or by sound waves. In such situations, we say energy is **dissipated**.

Supplement

Bungee jumping

What energy changes happen to a bungee jumper after jumping off the platform?

- The gravitational potential energy of the bungee jumper changes to kinetic energy as the jumper falls with the rope slack.
- Once the slack in the rope has been taken up, the rope slows the bungee jumper's fall. Most of the gravitational potential energy and kinetic energy of the jumper is changed into elastic energy in the rope as it stretches.
- After reaching the bottom, the rope pulls the jumper back up. As the jumper rises, **most of** the elastic energy of the rope changes back to gravitational potential energy and kinetic energy of the jumper.

The bungee jumper doesn't return to the same height as at the start. This is because some of the initial gravitational potential energy has been changed to heat energy as the rope stretched then shortened again.

Bungee jumping

PRACTICAL

Investigating a model bungee jump

1 You can try out the ideas about bungee jumping using the experiment shown in Figure 4.2.2.

2 Find out how much of the gravitational potential energy lost in the descent is regained at the end of the jump.

3 To do this, measure and compare the total height drop of the bungee jumper with the total height gain after the drop.

Figure 4.2.2 Testing a bungee jump

SUMMARY QUESTIONS

1 Complete the sentences below using words from the list.

electrical   gravitational potential   thermal

A person going up in a lift has _____ energy. The lift is driven by electric motors. Some of the _____ energy supplied to the motors is changed to _____ instead of _____ energy.

2 A ball dropped onto a trampoline returns to the same height after the rebound.

a Describe the energy change of the ball from the point of release to the top of the rebound.

b What can you say about the energy of the ball at the point of release compared with at the top of the rebound?

3 One exciting fairground ride acts like a giant catapult. The capsule which the 'rider' is strapped in is fired high into the sky by rubber straps. Explain the energy changes taking place in the ride.

KEY POINTS

1 Energy can be transformed from one form to another or transferred from one place to another.

2 Energy cannot be created or destroyed.

Supplement

LEARNING OUTCOMES

- Describe how electricity is generated in a small generator and in a power station
- Describe the energy transformations that take place when fossil fuels and biofuels are burnt in power stations
- Explain what is meant by efficiency
- Recall and use efficiency equations

STUDY TIP

Make sure you understand these energy transformations.

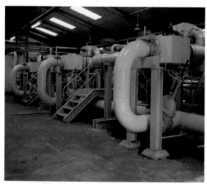

A petrol-powered electricity generator

In a gas-fired power station

Fuel for electricity

We can generate electricity by burning fuel and using the heat energy released to make an engine turn an electricity generator. Such fuels include:

- fossil fuels such as oil (including petrol), natural gas and coal
- biofuels such as wood, straw, methane gas and ethanol.

Note that nuclear fuels (e.g. uranium) release energy as a result of the nuclei of atoms disintegrating, not as a result of burning fuel.

The energy changes when electricity is generated as a result of burning fuel are:

chemical energy in the fuel

     ⟶ *heat energy (+ light energy)*

        ⟶ *kinetic energy of the engine and generator (+ sound energy + heat energy due to friction)*

          ⟶ *electrical energy*

Power from electricity

Electricity generators supply electrical energy continuously from the energy released when fuel burns. The **power** of an electricity generator is the rate at which it supplies electrical energy.

Petrol-powered electricity generators usually consist of a petrol engine that turns an electricity generator. In a petrol engine, petrol and air are ignited in each engine cylinder in sequence, keeping the 'drive' shaft, which is connected to the electricity generator, turning. Petrol generators are used when or where mains electricity is not available.

Power stations generate the electricity used by most people except those in remote locations.

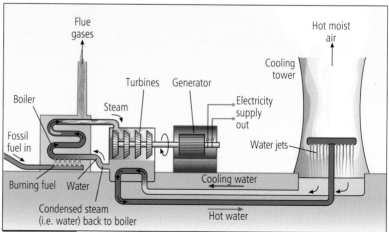

Figure 4.3.1   Inside a fossil fuel power station

- **In a coal- or oil-fired power station or a biofuel power station**, the burning fuel heats water in a boiler to produce steam (Figure 4.3.1). The steam drives a turbine that turns an electricity generator. After being used to drive the turbines, the steam condenses to water which is returned to the boiler. The condensed water is kept cool by cold water pumped through pipes that pass through the condensed water.
- **In a gas-fired power station**, natural gas is burned directly in a gas turbine engine. This produces a powerful jet of hot gases and air that drives the turbine. A gas-fired turbine can be switched on very quickly.

PRACTICAL

Electricity generator tests

1 A cycle dynamo is designed to light a torch bulb. The dynamo consists of a magnet that spins inside a coil of wire to which the torch bulb is connected. Turn the dynamo steadily and light the torch bulb.

2 A clockwork radio is designed to store energy in a clockwork spring which is used to drive a small electricity generator in the radio. Wind the clockwork spring and listen to the radio. Find out how long the electricity supply lasts for.

Energy and efficiency

Energy is measured in joules (J). As explained in detail in Topic 4.7, 1 J of energy is the energy transferred to an object of weight 1 N when it is raised by a vertical distance of 1 m.

When energy is supplied to a device, not all the energy supplied to it is used for the intended purpose. For example, when an electric motor is used to raise a weight, some of the electrical energy supplied to the motor is wasted as heat energy due to friction and as sound energy. The energy transferred by the motor to raise the load is **useful energy** because it is used for the intended purpose of raising the weight.

For any device, its efficiency is a measure of its useful energy output in terms of the energy supplied.

The energy supplied to it = the useful energy output of the device + the energy wasted by the device

Supplement

We can define the **efficiency** of a device as:

$$\frac{\text{the useful energy output of the device}}{\text{the energy supplied to the device}} \times 100\% \quad \text{or}$$

$$\frac{\text{the useful power output}}{\text{the power input}} \times 100\%$$

For example, suppose an electric motor used to raise a weight transfers 30 J of energy to the weight as gravitational potential energy for every 100 J of energy supplied to the motor. The other 70 J is transferred to the surroundings, mostly as heat energy and also as sound energy.

The percentage efficiency of the motor is 30% $\left(\frac{30\,\text{J}}{100\,\text{J}} \times 100\%\right)$.

KEY POINTS

1 A petrol generator consists of a petrol engine and an electricity generator.

2 Electricity generators in power stations are driven by turbines.

3 The unit of energy is the joule.

4 Efficiency is a measure of how much of the energy supplied to a device is usefully used.

SUMMARY QUESTIONS

1 Copy and complete the sentences below using words from the list.

**coal   gas   oil
uranium   wood**

a _____ and _____ are not fossil fuels.

b Power stations that use _____ as the fuel can be switched on very quickly.

c Steam is used to make the turbines rotate in a power station that uses coal, _____ or _____ as fuel.

2 a Name the device used to turn an electricity generator in:
   i   a petrol generator
   ii a gas-fired power station.

b State one advantage and one disadvantage of a gas-fired power station compared with a coal-fired power station.

c Draw an energy flow diagram to show the energy changes in a clockwork radio when the radio is on.

Nuclear energy

Supplement

LEARNING OUTCOMES

- Describe what nuclear fission is

- Describe how electricity is produced from nuclear fuel

- Describe what happens in nuclear fusion

- Explain why it is difficult to sustain nuclear fusion in a fusion reactor

STUDY TIP

It is very important to learn the facts about the structure of an atom.

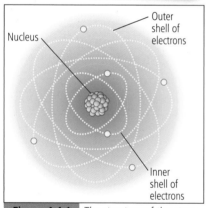

Figure 4.4.1	The structure of the atom

Inside the atom

Figure 4.4.1 shows the particles inside the atom.

- Every atom contains a positively charged nucleus surrounded by electrons which are negatively charged.
- The nucleus is composed of neutrons and protons.
- Atoms of the same element can have different numbers of neutrons in the nucleus.

Nuclear fission

The fuel in a nuclear power station is uranium. It releases about 10 000 times as much energy per kilogram as fossil fuel or biofuel. The uranium fuel is contained in sealed cans in the core of the reactor.

The nucleus of a uranium atom is unstable and can split in two when a neutron from outside the atom collides with it. This process is called **nuclear fission**. Energy is released in the process.

Nuclear power stations do not produce greenhouse gases whereas fossil fuel power stations do. However, they produce radioactive waste which needs to be stored safely for many years.

Inside a fission reactor

When a uranium nucleus undergoes fission, two or three neutrons may be released, as well as energy. These neutrons may cause other uranium nuclei to undergo fission. When this happens, a chain reaction can occur in the core of the reactor.

As there are lots of uranium atoms in the core, it becomes very hot. The thermal energy of the core is taken away by a fluid (called the 'coolant') that is pumped through the core. The coolant is very hot when it leaves the core. It flows through a pipe to a 'heat exchanger' then back to the reactor core. The thermal energy of the coolant is used to turn water into steam in the heat exchanger. The steam drives turbines which turn electricity generators (Figure 4.4.2).

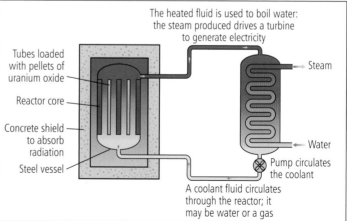

Figure 4.4.2	Left: a nuclear power station. Right: components of a nuclear reactor

Nuclear fusion

The Sun and the stars release energy as a result of fusing small nuclei, such as hydrogen, to form larger nuclei. Two small nuclei release energy when they are fused together to form a single larger nucleus (Figure 4.4.3). The process is called **nuclear fusion**.

STUDY TIP

'Fission' means splitting. Don't confuse nuclear fission with nuclear fusion.

Supplement

Fusion reactors

Fusion reactors release energy by fusing hydrogen nuclei and other light nuclei. The fuel needed, such as hydrogen, is available in abundance and the fusion products are not radioactive. However, there are enormous technical difficulties in sustaining fusion in a fusion reactor. A collection of unbound nuclei and electrons, as in a fusion reactor, is called a plasma. The plasma must be heated to very high temperatures before any of the nuclei will fuse with each other. This is because two nuclei approaching each other will repel each other due to their positive charge. If the nuclei are moving fast enough, they can overcome the force of repulsion and fuse together.

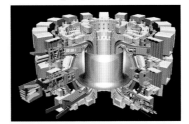

An experimental fusion reactor

Figure 4.4.3 A fusion reaction. The fusion of two nuclei to form a new nucleus which is lighter than iron's generally releases energy

In a fusion reactor:

• the plasma is heated by passing a very large electric current through it

• the plasma is contained by a magnetic field so it doesn't touch the reactor walls. If it did, it would go cold and fusion would stop.

Scientists have not yet developed a successful fusion reactor which would release more energy than it uses to heat the plasma. At the present time, scientists working on experimental fusion reactors are able to do this by fusing hydrogen nuclei to form helium nuclei – but only for a few minutes!

DID YOU KNOW?

Water contains lots of hydrogen atoms. A glass of water could provide the same amount of energy as a tanker full of petrol – if we could make a fusion reactor here on the Earth.

SUMMARY QUESTIONS

1 Copy and complete the sentences below using the following words.

large   small   stable

a   When nuclear fission occurs, a _____ nucleus splits into two _____ nuclei.

b   Energy is released in nuclear fusion if the product nucleus is not as _____ as an iron nucleus.

c   When two _____ nuclei moving at high speed collide, they form a _____ nucleus.

Supplement

2 a   Explain what is meant by a chain reaction in a nuclear fission reactor.

b   Why does the plasma in a fusion reactor need to be very hot?

KEY POINTS

1 Energy is released in nuclear fission.

Energy is also released in nuclear fusion.

2 Nuclear fission occurs when a uranium nucleus splits as a result of being struck by a neutron.

3 Nuclear fusion occurs when two nuclei are forced close enough together so they form a single larger nucleus.

Supplement

LEARNING OUTCOMES

- Describe how the wind can be used to generate electricity
- Describe how falling water can be used to generate electricity
- Describe how waves and tides are used to generate electricity

Wind energy

A wind turbine is an electricity generator at the top of a narrow tower. The generator is driven by the force of the wind on its blades. The power generated increases as the wind speed increases. A wind farm consisting of 40 wind turbines could generate enough electricity to supply over 10 000 homes.

Wind turbines do not emit greenhouse gases and do not cause pollution. However, if the wind does not blow, they cannot produce electricity – so they are unreliable.

A wind farm

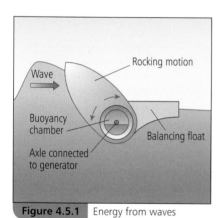

Figure 4.5.1 Energy from waves

Wave energy

A wave generator at a coastal site uses the motion of waves to make a floating section move up and down (Figure 4.5.1). This motion drives a turbine which turns a generator. A cable between the generator and the shore delivers electricity to local users or via a network of cables (referred to as the grid system) to distant users.

Wave generators need to withstand storms and they don't produce a constant supply of electricity. Furthermore, lots of cables (and buildings) would be needed along the coast to connect the wave generators to the electricity grid. This would spoil the views of the coastline. In addition, tidal flow patterns might be changed, affecting the habitats of marine life and birds.

Hydroelectricity

Hydroelectricity is generated when rain water collected in an uphill reservoir flows downhill. The water flow drives turbines that turn electricity generators at the foot of the hill. Large hydroelectric power stations can supply enough electricity for a city. The 'Three Gorges' hydroelectric scheme in China traps fast-flowing river water behind a massive dam and channels the water through generators that supply electricity to millions of homes.

Hydroelectric schemes do not produce greenhouse gases, although they can have a considerable environmental impact if they require the construction of large dams and the flooding of large areas of land.

A hydroelectric scheme

Tidal energy

A tidal power station in an estuary traps each high tide behind a barrage. The high tide is then released into the sea through turbines that drive generators in the barrage.

Tidal energy is more reliable than wind energy because ocean tides occur roughly twice every day. They happen because the Moon's gravity pulls on the Earth's oceans.

The table below shows that a tidal power station can generate much more power than a wind turbine or a hydroelectric power station.

Power station (or devices)	Typical output	Location	Cost[a] per MW h
Coal-fired	1000 MW	Any	1
Hydroelectric	500 MW	Upland	0.8
Nuclear	5000 MW	Coastal	1
Solar cells	1 kW per m²	Any	1.3
Tidal	2000 MW	Estuary	2
Wave generators	20 MW per km	Coast	0.9
Wind turbines	2 MW per turbine	Windy site	0.8[b] 2.1[c]

Notes   **1**   *1 MW = 1 million watts = 1 million joules per second*

       2 a   *Costs are given relative to coal;* **b** *onshore;* **c** *offshore*

The energy sources described in this topic and solar energy in the next topic are called renewable energy sources because they never run out – unlike non-renewable sources such as fossil fuels, nuclear fuels and geothermal energy sources.

DID YOU KNOW?

Pumped storage schemes pump water uphill and store it in uphill reservoirs. This is useful when the demand for electricity is low. When demand goes up again, the water is allowed to flow downhill and generate electricity.

A tidal power station at a suitable location can generate as much power as a large fossil-fuel power station or a nuclear power station, enough to supply a city. An estuary that becomes narrower as you move 'upriver' away from the open sea is a good site for a tidal power station. This is because it 'funnels' the incoming tide and makes it higher than elsewhere.

SUMMARY QUESTIONS

1 Copy and complete the following sentences using words from the list below.

   hydroelectric   tidal   wave   wind

   a   _____ energy does not need water.

   b   _____ energy does not need energy from the Sun.

   c   _____ energy is obtained from water running downhill.

   d   _____ energy is obtained from water moving up and down.

2 a   Use the table above for this question.

   i   How many wind turbines would give the same total output as a tidal power station?

   ii   How many kilometres of wave generators would give the same total output as a hydroelectric power station?

   b i   How many 2 MW wind turbines would give the same power output as a 5000 MW nuclear power station?

   ii   State one advantage and one disadvantage of installing wind turbines instead of building a nuclear power station.

KEY POINTS

1 A wind turbine is an electricity generator on top of a tall tower.

2 A wave generator is a floating generator turned by the waves.

3 Hydroelectricity generators are turned by water running downhill.

4 A tidal power station traps each high tide and uses it to turn generators.

Energy from the Sun and the Earth

LEARNING OUTCOMES

- Describe a solar cell panel and explain what we use it for
- Describe a solar heating panel and explain what we use it for
- Explain what geothermal energy is and how we use it to generate electricity

A solar powered vehicle

DID YOU KNOW?

The Sun is the source of energy for all our energy resources except nuclear, tidal and geothermal energy. Fossil fuels formed from the remains of plants and animals that grew using solar energy long ago. The Sun heats our atmosphere, creating winds, waves and rain as a result.

STUDY TIP

Make sure that you understand the relevance of the BLACK cover and the COPPER pipes in a solar heating panel.

Solar energy

Solar radiation transfers energy to you from the Sun – sometimes more than you want if you get sunburnt. We can use it to generate electricity using **solar cells** and we can also use it to heat water directly in **solar heating panels**.

1 **Solar cells** in use now convert about 15% of the solar energy they absorb into electrical energy. We connect them together to make solar cell panels.

   - They are useful where only small amounts of electricity are needed (e.g. watches and calculators) or in remote locations (e.g. satellites).
   - They are very expensive to buy even though they cost nothing to run.
   - You need lots of them to generate enough power to be useful – and plenty of sunshine!
   - They don't work at night and aren't very effective on a cloudy day.

2 **A solar heating panel** heats water flowing through it. Even on a cloudy day, a solar heating panel on a house roof can supply plenty of hot water (Figure 4.6.1).

   - The panel has a black cover which absorbs sunlight better than other colours.
   - The back of the panel is insulated so water flowing through the panel doesn't lose heat energy.
   - The panel is filled with oil so it heats up faster than water.
   - Water passes through copper pipes in the oil. Copper is a good conductor of heat so heat energy is readily transferred from the hot oil to the water in the pipes.
   - Cold water enters the pipes and is heated by the oil. The hot water from the outflow pipe is collected in an insulated tank.

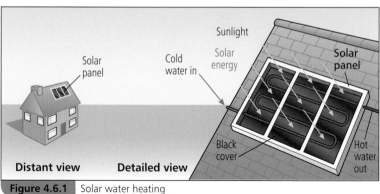

Figure 4.6.1 Solar water heating

Investigating solar cells

1 Use a solar cell to drive a small motor (Figure 4.6.2).

2 Observe the change of motor speed when you use a card to cover the solar cell partially then totally.

3 Carry out the test on a sunny day and on a cloudy day to see what difference cloud cover makes.

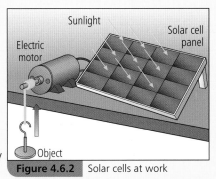

Figure 4.6.2 Solar cells at work

Geothermal energy

Geothermal energy is released by radioactive substances deep within the Earth.

• The energy released by these substances heats the surrounding matter.

• As a result, heat energy transfers outward towards the Earth's surface.

• Rocks below the surface become very hot because they gain thermal energy from the flow of heat energy.

Geothermal power stations are built where there are hot rocks deep below the surface. Water is pumped down to these rocks to produce steam. Electricity turbines at ground level are driven by the steam.

Geothermal power stations do not need energy from the Sun because the energy they use comes from radioactive substances deep within the Earth.

The Gobi desert is one of the most remote regions on Earth. Yet people who live there can watch TV programmes just like you can. All they need is a solar panel and satellite TV.

Geothermal energy

Frying Pan Lake in New Zealand gets its scalding water from deep underground. The lake is in a large crater formed after a volcanic eruption in 1886.

1 Copy and complete the following sentences using words from the list.

   geothermal energy   solar energy   radiation radioactivity

   a The best energy resource to use in a calculator is _____.

   b _____ inside the Earth releases _____ energy.

   c _____ from the Sun generates electricity in a solar cell.

2 a A satellite in space uses a solar cell panel for electricity. The satellite carries batteries that are charged by electricity from the solar cell panels. Why are batteries carried as well as solar cell panels?

   b If the water stopped flowing through a solar heating panel, what would happen to the temperature of the water in the panel? Give a reason for your answer.

1 Solar energy can be converted into electricity using **solar cells** or used to heat water directly in **solar heating panels.**

2 Geothermal energy is released by radioactive substances deep inside the Earth.

3 The Sun is the source of energy for all energy resources except geothermal energy, nuclear energy and tidal energy.

Supplement

Supplement

LEARNING OUTCOMES

- Recognise that when a force moves an object, increasing the force or the distance moved increases the work done

- Recall and use equations to calculate:
  - the work done by a force
  - the change of gravitational potential energy when an object is raised or lowered
  - the kinetic energy of a moving object

STUDY TIP

This is a simple equation but one that students often seem to forget!

Working out

STUDY TIP

This is another important equation to learn.

Working out

In a fitness centre or a gym, you have to work hard to keep fit. Raising weights and using a running machine are just two ways to keep fit. Whichever way you choose to keep fit, you have to apply a force to move something. The work you do causes transfer of energy. For example, if you raise an object and increase its gravitational potential energy by 20 J, the work you do on the object is 20 J.

Work done = energy transferred

The work done by a force depends on the force and the distance moved and is defined as the force (in newtons) × distance moved in the direction of the force (in metres)

Thus the work done, ΔW, (in joules) when a force F moves an object by a distance d in the direction of the force is given by the equation:

$$\Delta W = F \times d$$

where the force is in newtons and the distance moved is in metres.

Gravitational potential energy

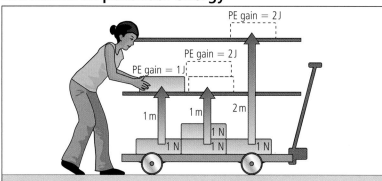

Figure 4.7.1 Using joules

Gravitational potential energy is the energy of an object due to its position. Consider an object of mass m raised through a vertical distance h:

- the weight of the object = mg, where g is the gravitational field strength at the Earth's surface,
- the force needed to raise the object is equal and opposite to its weight.

Therefore:

the gain of gravitational potential energy of the object = work done on the object to raise it

= force × distance moved = $(mg)h = mgh$

In general, when an object of mass m is raised or lowered by a vertical distance h:

its change of gravitational potential energy = mgh

Kinetic energy

The kinetic energy of an object is its energy due to its motion. It can be shown that the kinetic energy of an object of mass m moving at speed v is given by the equation:

kinetic energy $= \frac{1}{2}mv^2$, where m = the mass of the object in kilograms
(in joules, J)   and v = the speed of the object in m/s.

STUDY TIP

...and another important equation!

WORKED EXAMPLE

Calculate the kinetic energy of a vehicle of mass 500 kg moving at a speed of 12 m/s.

Solution

Kinetic energy $= \frac{1}{2}mv^2 = 0.5 \times 500\,\text{kg} \times (12\,\text{m/s})^2 = 36\,000\,\text{J}$

A ride on a roller coaster at a fairground causes rapid changes in the kinetic energy and gravitational potential energy of the riders and the train. Suppose the highest point of the track is followed by a steep descent. In the descent, the train loses gravitational potential energy and gains kinetic energy. Assuming the loss of gravitational potential energy of the train is transformed to kinetic energy, the speed, v, at the bottom of the descent is given by:

$$\frac{1}{2}mv^2 = mgh$$

where h is the vertical distance between the highest point of the track and the bottom of the descent.

In practice, due to air resistance and friction, the speed given by the above equation is not reached.

WORKED EXAMPLE

A 20 N weight is raised through a height of 0.4 m. Calculate:
a the work done, **b** the gain of gravitational potential energy of the object.

Solution

a The force needed to lift the weight $= 20\,\text{N}$

Work done
= force × distance moved in the direction of the force
$= 20\,\text{N} \times 0.4\,\text{m} = 8.0\,\text{J}$.

b Gain of gravitational potential energy = work done = 8.0 J.

DID YOU KNOW?

The top speed on a roller coaster of height 60 m is about 35 m/s (which is about 125 km/h).

SUMMARY QUESTIONS

Use $g = 10\,\text{N/kg}$

1 a Calculate the work done when:

  i a force of 20 N makes an object move 4.8 m in the direction of the force,

  ii an object of weight 80 N is raised through a height of 1.2 m.

2 a An object of weight 2.0 N fired vertically upwards from a catapult reaches a maximum height of 5.0 m. Calculate:

  i the gain of gravitational potential energy of the object,

  ii the kinetic energy of the object when it left the catapult.

 b A roller coaster train of mass 1800 kg descends through a vertical distance of 50 m from rest to the bottom of the track. Calculate:

  i the loss of gravitational potential energy of the train at the bottom of the descent,

  ii the maximum possible speed of the train at the bottom of the descent.

KEY POINTS

1 Work done ΔW by a force depends on the force F and distance d moved in the direction of the force:

$$\Delta W = Fd$$

2 For an object of mass m, its change of gravitational potential energy $= mgh$ where h = vertical distance moved,

its kinetic energy $= \frac{1}{2}mv^2$, where v = its speed.

Power

Supplement

LEARNING OUTCOMES

- Recognise that power is rate of transfer of energy
- Recall that power is measured in watts
- Recall and use the equation:

$$\text{power} = \frac{\text{energy transferred}}{\text{time taken}}$$

Rocket power for the first moon landing

STUDY TIP

Note that all types of power are measured in watts (W).

Powerful machines

When you use a lift to go up, a powerful electric motor pulls you and the lift up. The work done by the lift motor transfers energy from electricity to gravitational potential energy. Thermal energy is also produced and some sound energy.

- The work done per second by the motor is the output **power** of the motor.
- The more powerful the lift motor is, the faster it takes you up.

We measure the power of an appliance in watts (W) or kilowatts (kW) or megawatts (millions of watts) MW. One watt is a rate of transfer of energy of 1 joule per second (J/s).

For example:

- a 5 W electric torch would transfer 5 J every second as light energy and heat energy to its surroundings.
- a lift motor with an output power of 6000 W would transfer 6000 J to the lift as gravitational potential energy every second.

Here are typical values of power levels for different energy transfer 'mechanisms':

- A torch   1 W
- An electric light bulb   100 W
- An electric cooker   10 000 W
  = 10 kW (where 1 kW = 1000 watts)
- A railway engine   1 000 000 W
  = 1 megawatt (MW) = 1 million watts
- A Saturn V rocket   100 MW
- A very large power station   10 000 MW
- The Sun   100 000 000 000 000 000 000 MW

Energy and power

Supplement

Machines are labour-saving devices that do work for us. The faster a machine can do work, the more powerful it is.

Whenever a machine does work on an object, energy is transferred to the object. The useful energy transferred is equal to the work done.

The output power of a machine is the rate at which it does work. This is the same as the rate at which it transfers useful energy.

$$\textbf{power } P \textbf{ (in watts)} = \frac{\textbf{work done (in joules)}}{\textbf{time taken (in seconds)}}$$

$$= \frac{\textbf{useful energy transferred (in joules)}}{\textbf{time taken (in seconds)}}$$

If energy E is transferred in time t,

$$\text{power } P = \frac{E}{t}$$

A crane at work

WORKED EXAMPLE

1 A crane lifts an object of weight 4000 N through a vertical distance of 2.5 m in 5.0 s. Calculate:

  a the gain of gravitational potential energy of the object
  b the output power of the crane.

Solution

  a Gain of gravitational potential
     energy = mgh = 4000 N × 2.5 m = 10 000 J
  b Work done by the crane = gain of gravitational potential
     energy = 10 000 J

$$\text{Output power} = \frac{\text{work done}}{\text{time taken}} = \frac{10\,000\,\text{J}}{5.0\,\text{s}} = 2000\,\text{W}$$

PRACTICAL

Measure your personal power

1 Ask a friend to time how long it takes you to step on and off a suitable platform ten times. If you are medically unfit, you do the timing and your friend can do the steps.

2 For each step, your increase of gravitational potential energy = your weight × the height of the platform. This is the work you do each step.

3 Calculate the work you do for ten steps. Your personal power = work done for ten steps ÷ time taken.

SUMMARY QUESTIONS

1 a Which is more powerful?
    i   a torch bulb or a mains filament lamp?
    ii a 3 kW electric kettle or a 10 000 W electric cooker?

  b There are about 2 million homes in a certain city. If a 3 kW electric kettle was switched on in 1 in 10 homes in the city at the same time, how much power would need to be supplied?

2 An electric motor raises an object of weight of 200 N through a height of 4.0 m in 5.0 s. Calculate:

  a the gain of gravitational potential energy of the object,
  b the work done by the motor on the object,
  c the output power of the motor.

DID YOU KNOW?

Muscle power

How powerful is a weight lifter? A 30 kg dumbell has a weight of 300 N. Raising it by 1 m would give it 300 J of gravitational potential energy. A weight lifter could lift it in about 0.5 s. The rate of transfer of energy would be 600 J/s (= 300 J/0.5 s). So a weight lifter's output power would be about 600 W in total!

Figure 4.8.1 Steps

KEY POINTS

The unit of power is the watt (W), equal to 1 J/s.

1 kilowatt = 1000 watts

$$\text{power (in watts)} = \frac{\text{energy transferred}}{\text{time taken}}$$

1 Describe the following forms of energy:

(a) kinetic energy

(b) gravitational potential energy

(c) nuclear energy

(d) chemical energy.

2 The school laboratory has a small electric motor that runs from two 1.5V cells. A pulley wheel is mounted on the axle and this can be connected to turn another pulley wheel that is mounted on the axle of a small dynamo (dc generator). The dynamo is connected to a lamp. When the circuit is switched on the motor turns. This makes the dynamo turn and this lights the lamp.

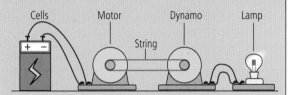

(a) Show with an energy flow diagram the energy changes at each stage from the cells to the lamp.

(b) Explain why the lamp will be lit more brightly if connected directly to the cells.

3 An electric motor is used to lift a load. When 6000J of electrical energy are supplied to the motor the load gains 2400J of gravitational potential energy.

(a) Calculate the amount of energy wasted by the motor.

(b) Suggest in which form(s) the energy is wasted.

4 (a) Suggest one advantage and one disadvantage of each of the following types of energy resource when used as a method of electricity generation.

(i) Tidal energy

(ii) Wave energy

(iii) Geothermal energy

(iv) Solar energy

(b) Choose two of the above methods of electricity generation and describe how the necessary energy transformation is achieved.

5 Using the information in question **3**:

Calculate the efficiency of the motor.

1 The unit of energy is

  A J   **C** N

  B kg   **D** W

(Paper 1/2)

2 A machine is used to raise a 60 kg load from the ground to a height of 5 m. The gain in gravitational energy of the load is

  A 12 J   **C** 300 J

  B 120 J   **D** 3000 J

(Paper 2)

3 Which of the following is NOT a statement of the Principle of Conservation of Energy?

  A Energy cannot be created or destroyed

  B Energy should not be wasted

  C The total amount of energy before and after a change is the same

  D The total amount of energy in the universe is constant

(Paper 1/2)

4 Which of these is NOT a form of energy?

  A gravity   **C** light

  B kinetic   **D** nuclear

(Paper 1/2)

5 A machine uses 120 kJ of energy in 4 min. The power of the machine is

  A 0.5 kW   **C** 480 kW

  B 30 kW   **D** 28 800 kW

(Paper 1/2)

6 A student has a mass of 65 kg. He climbs up a flight of 12 stairs. Each stair has a height of 16 cm. The time taken to reach the top is 2.2 s ($g = 10$ N/kg).

(a) Calculate the student's weight.

(b) Calculate the gravitational energy gained by the student between the bottom and the top of the flight of stairs.

(c) Calculate the student's power.

16 cm

Supplement

6 (a) Match the form of energy in the left-hand column with its description in the right-hand column.

Chemical energy	energy transfer from a hot object to a cold object
Electrical energy	energy of an object due to its motion
Gravitational potential energy	energy released in a substance when chemical reactions take place
Heat energy	energy released when the nucleus of an atom splits
Kinetic energy	energy transferred by an electric current
Nuclear energy	energy of an object due to its position

[3]

(Paper 3)

(b) Copy and complete the energy transformation flow diagrams by filling in the missing words.

An electric torch.

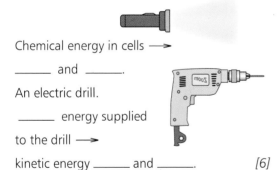

Chemical energy in cells ⟶

_____ and _____.

An electric drill.

_____ energy supplied

to the drill ⟶

kinetic energy _____ and _____.   [6]

7 (a) Suggest ONE advantage and ONE disadvantage of generating electrical power using the following methods:

(i) Wind turbine

(ii) A hydroelectric scheme.   [4]

(b) Copy and complete these sentences:

(i) The wind turbine transforms _____ to _____ .

(ii) The hydroelectric scheme transforms _____ stored in the water behind the dam to _____ as the water falls. This is then transformed to _____ (plus some heat and sound) in the generator.   [5]

(Paper 3)

8 A model steam engine raises a 1.5 kg load from the floor to the laboratory bench with a height of 0.80 m in a time of 4 s.

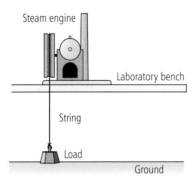

(a) (i) Calculate the work done in raising the load. (Assume that $g = 10 \text{ m/s}^2$.)

(ii) Calculate the power of the model engine.

(iii) State the value of the gravitational potential energy of the load when it is on the bench.   [5]

(b) A second model steam engine is available.

(i) Using the same load and laboratory bench as in part **a**, describe how you would test this steam engine to determine whether it is more powerful, less powerful or of the same power as the first model steam engine.

(ii) State how your results would compare with those for the first steam engine if the second were: 1) more powerful 2) less powerful 3) of the same power.   [5]

(c) A student releases the load from the laboratory bench so that it falls to the ground. A timing device measures its velocity immediately before it hits the ground. The velocity is 3 m/s.

(i) Calculate the kinetic energy of the load as it hits the ground.

(ii) Calculate the difference between the kinetic energy of the load as it hits the ground and its gravitational energy when it is on the laboratory bench.

(iii) The student expected the kinetic energy and gravitational energy to be the same. Suggest a reason for the two values being different.   [5]

(Paper 4)

5 Pressure

5.1

Under pressure

LEARNING OUTCOMES

- Recall that pressure is force per unit area
- Recognise that a larger area for a given force reduces the pressure of the force
- Explain applications where the area is an important design consideration
- Recall and use the equation:

$$\text{Pressure} = \frac{\text{force}}{\text{area}}$$

STUDY TIP

Learn to use the correct words. For example a SMALL force over a LARGE area produces a LOW pressure.

Caterpillar tracks

What is pressure?

If you stand barefoot on a sharp object, you will find out about pressure in a very painful way. All your weight acts on the tip of the object so there is huge pressure on your foot at the area of contact.

Pressure is caused when objects exert forces on each other. The pressure caused by any force depends on the area of contact where the force acts, as well as on the size of the force.

Caterpillar tracks fitted to vehicles are essential on sandy or muddy ground or on snow-covered ground. The reason is that the contact area of the tracks on the ground is much larger than it would be if the vehicle had wheels instead. The tracks therefore reduce the pressure of the vehicle on the ground as its weight is spread over a much larger contact area. It is therefore less likely to sink into the sand, mud or snow.

Pressure is defined as force per unit area. The unit of pressure is the pascal (Pa) which is equal to one newton per square metre (N/m²).

For a force F acting evenly on a surface of area A, at right angles to the surface, the pressure p on the surface is given by the equation:

$$p = \frac{F}{A}$$

Note: Rearranging this equation gives $F = p \times A$ or $A = \frac{F}{p}$.

WORKED EXAMPLE

A caterpillar vehicle of weight 12 000 N is fitted with tracks that have an area of 3.0 m² in contact with the ground. Calculate the pressure of the vehicle on the ground.

Solution

$$\text{Pressure} = \frac{\text{force}}{\text{area}} = \frac{12\,000\,\text{N}}{3.0\,\text{m}^2} = 4000\,\text{Pa}$$

PRACTICAL

Measure your foot pressure

1 Draw around your shoes using centimetre squared paper. Count the number of centimetre squares in each footprint (ignoring any square that is less than half-filled) to find the area of contact in square centimetres. Convert this to square metres using the conversion 1 m² = 10 000 cm².

2 Use suitable scales (e.g. bathroom scales) to measure your weight. If the scales read mass in kilograms, find your weight in newtons using $g = 10$ N/kg.

Work out your pressure using: $\text{pressure} = \dfrac{\text{weight}}{\text{area}}$

In hospital

A sharp knife cuts more easily than a blunt knife. Surgical knives used in operating theatres need to be very sharp. A sharp knife has a much smaller area of contact when it is used than a blunt knife. So the pressure (= force ÷ contact area) of a sharp knife is much greater than the pressure of a blunt knife for the same force. Therefore the same force applied to a sharp knife has a much greater effect than it would with a blunt knife.

Bed sores are a problem for patients confined to bed for long periods of time. Such sores occur where the body presses on the bed for a long time. The skin in contact with the bed is not as tough as the skin under your feet. As a result, the skin in the contact area is easily rubbed away, causing bed sores.

PRACTICAL

A pressure test

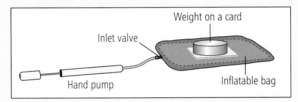

Weight on a card

Inlet valve

Hand pump

Inflatable bag

1   Use the arrangement shown to measure the pressure inside an inflatable bag.

2   With the bag deflated, place a weight on a card on the bag. Use the hand pump to inflate the bag until the card and weight are raised. The pressure of the air in the bag is then equal to the pressure of the weight on the card.

3   Measure the area of the card and calculate the pressure of the air in the bag (= weight in newtons ÷ area of the card in square metres).

KEY POINTS

1   Pressure is force per unit area.

2   The unit of pressure is the pascal (Pa) which is equal to $1\,N/m^2$.

3   For a force F acting at right angles on an area A, the pressure $p = \dfrac{F}{A}$.

SUMMARY QUESTIONS

1 Explain each of the following:

   a   When you do a handstand, the pressure on your hands is greater than the pressure on your feet when you stand upright.

   b   Snowshoes like those shown in the photograph are useful for walking across soft snow.

2 A rectangular concrete paving slab of weight 1200 N has sides of length 0.60 m and 0.40 m and a thickness of 0.05 m. Calculate the pressure of the paving slab on the ground when it is:

   a   laid flat on a bed of sand

   b   standing upright on its short side.

LEARNING OUTCOMES

- Recognise that pressure can be transmitted through a fluid
- Explain how a hydraulic system works
- Recall that the force exerted by a hydraulic system depends on the pressure and the area of the cylinder that exerts the force

a

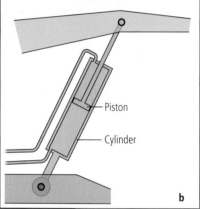

- Piston

- Cylinder

b

Figure 5.2.1 (a) A mechanical digger, (b) A hydraulic system

A remote-controlled robot handling a dangerous object

Mechanical diggers are used to remove large quantities of earth, for example when soil has to be removed above an underground pipe to reach the pipe. The 'grab' of the digger is operated by a **hydraulic pressure** system. The hydraulic system of a machine is its 'muscle power'.

In the hydraulic systems shown in Figure 5.2.1, oil is pumped into the upper or lower part of the cylinder to make the piston move in or out of the cylinder.

Robots use hydraulics for muscle power. Robotic machines in factories are fixed machines that operate such things as welding gear or paint sprays on assembly lines. Robot muscles use compressed air or oil and they work non-stop without the need for a human operator. Remote-controlled robots are used for dangerous tasks such as bomb disposal and handling suspect packages.

Vehicle brakes use pressure. When the driver presses on the brake pedal of the car, pressure is exerted on the brake fluid in the main cylinder. This pressure is transmitted along the brake pipes to wider cylinders at the wheels. The fluid pressure forces the piston in each wheel cylinder to push the brake disc pads on to the wheel disc.

Power-assisted brakes fitted to heavy goods vehicles and coaches use compressed air. When the driver applies the brakes, compressed air at very high pressure is released to push on the piston in the main cylinder. The compressed air is used instead of the brake pedal to exert the force on the main cylinder. This is why such vehicles hiss when the brakes are released.

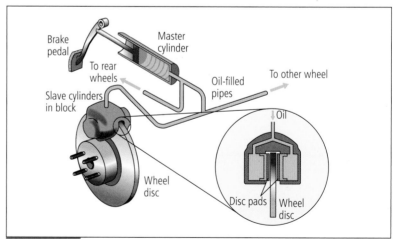

Figure 5.2.2 Disc brakes in a car

A hydraulic car jack can be used to lift a car, as shown in Figure 5.2.3. When the handle is pressed down, the piston in the narrow cylinder is forced into the oil-filled cylinder. Oil is forced out of this cylinder, through the pipe and into a wider cylinder. The pressure of the oil on the piston in the wider cylinder forces this piston outwards which forces the pivoted lever to raise the car.

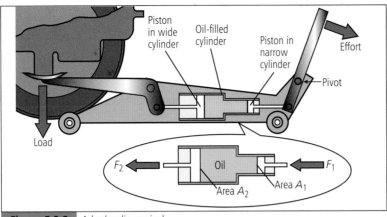

Figure 5.2.3 A hydraulic car jack

Supplement

In Figure 5.2.3:

- The force F_1 created by moving the lever acts on the narrow cylinder and creates a pressure $p = \dfrac{F_1}{A_1}$ on the fluid in the cylinder, where A_1 is the area of the narrow cylinder.
- This pressure is transmitted through the fluid in the pipe to the wider cylinder.
- The force on the larger piston, $F_2 = p \times A_2$ where A_2 is the area of the wide cylinder. Therefore $F_2 = \dfrac{F_1}{A_1} \times A_2$

The force F_2 is therefore much greater than F_1 because area A_2 is much greater than area A_1.

KEY POINTS

1 Pressure can be transmitted through a fluid.

2 A hydraulic system uses the pressure in a fluid to exert a force.

3 The force exerted by a hydraulic system depends on the pressure and the area of the cylinder that exerts the force.

SUMMARY QUESTIONS

1 a Write down as many machines as you can think of that are operated hydraulically.

b The photograph shows the arm of a mechanical digger. It is controlled by three hydraulic pistons called 'rams', labelled X, Y and Z.

i Explain why the arm is raised when compressed air is released into ram X so it extends.

ii State and explain what happens to the ' bucket' on the end of the arm when rams Y and Z are both extended.

2 The hydraulic lift shown in Figure 5.2.4 is used to raise a vehicle so its underside can be inspected.

The lift has four pistons, each of area $0.01\,\text{m}^2$ to lift the platform. The pressure in the system must not be greater than $5.0 \times 10^5\,\text{Pa}$. The platform weight is $2000\,\text{N}$. Calculate the maximum load that can be lifted on the platform.

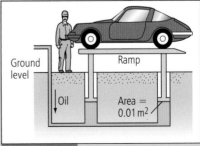

Figure 5.2.4

Pressure in a liquid at rest

LEARNING OUTCOMES

- Recall that the pressure in a liquid increases with increase of depth
- Recall that the pressure in a liquid depends on the density of the liquid
- Recall and use the equation for pressure due to a liquid column

Using a snorkel tube

STUDY TIP

This is a very simple experiment with an important result.

An underwater swimmer using a snorkel tube can breathe safely provided the top of the tube is above the water. However, a deep-sea diver could not breathe through a very long snorkel tube because the pressure of the water increases with depth. At depths of more than a few metres, the diver's chest muscles would not be strong enough to expand his or her chest muscles against the water pressure on the body.

The pressure of a liquid increases with depth. Figure 5.3.1a shows water jets from the holes down the side of an open plastic bottle filled with water. The further the hole is below the level of water in the bottle, the greater the pressure of the jet.

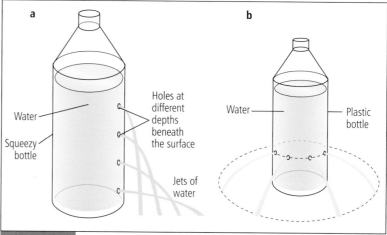

a

Water

Squeezy bottle

Holes at different depths beneath the surface

Jets of water

b

Water

Plastic bottle

Figure 5.3.1 Pressure in a liquid at rest: (a) pressure increases with depth, (b) same pressure at the same depth

The pressure along a horizontal line in a liquid is constant. This can be demonstrated by making several holes around the bottle at the same depth, as in Figure 5.3.1b. The jets from these holes are at the same pressure.

DID YOU KNOW?

Figure 5.3.2 shows several containers joined to a sealed pipe. If a liquid is poured into one of the containers, some of the liquid flows into the other containers. The flow stops when the level of the liquid in each container is the same. This is because the pressure in each container along the same horizontal line has become equal. The unit of pressure, the pascal (Pa), is named after the 17th century French scientist, Blaise Pascal, who first demonstrated the above effect and made many other discoveries about pressure.

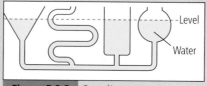

Level

Water

Figure 5.3.2 Pascal's vases

The pressure in a liquid depends on the density of the liquid.
Suppose water is poured into one side of a U-shaped tube, then oil is carefully poured into the other side, as in Figure 5.3.3. When the liquids settle, the oil level is higher than the water level on the other side. This is because oil is less dense than water so a greater depth of oil is needed to create the same pressure.

Supplement

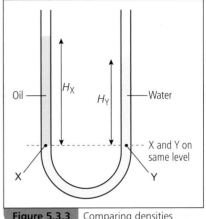

Figure 5.3.3 Comparing densities

The pressure of a liquid column

Consider the column of liquid in the container shown in Figure 5.3.4. The pressure caused by the liquid column on the bottom of the container is due to the weight of the liquid.

For a column of height h and area of cross-section A, the volume of liquid in the container $= hA$.

The mass of liquid $=$ its volume $\times$ density of the liquid $= hA\rho$, where ρ is the density of the liquid.

The weight of the liquid $=$ mass $\times g = hA\rho \times g$, where the gravitational field strength $g = 10\,$N/kg.

The pressure p at the base of the liquid column

$$= \frac{\text{weight}}{\text{area of cross-section}} = \frac{hA\rho g}{A}.$$

Therefore, cancelling A gives: $\boldsymbol{p = h\rho g}$

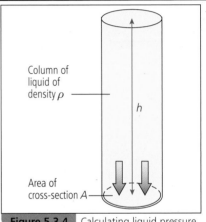

Figure 5.3.4 Calculating liquid pressure

Column of liquid of density ρ

h

Area of cross-section A

WORKED EXAMPLE

Use $g = 10\,$N/kg

Calculate the pressure due to sea water of density $1050\,$kg/m³ at a depth in the sea of $200\,$m.

Solution

Pressure $p = h\rho g = 200\,\text{m} \times 1050\,\text{kg/m}^3 \times 10 = 2.1 \times 10^6\,$Pa.

SUMMARY QUESTIONS

Use $g = 10\,$N/kg

1 a Explain why the wall of a dam needs to be thicker at the base than at the top.

  b A water tank at the top of a tall building supplies water to taps in the building. Explain why the pressure of the water from a tap on the ground floor is greater than the pressure from a tap on a higher floor.

Supplement

2 A sink plug has an area of $0.0006\,$m². It is used to block the outlet of a sink filled with water to a depth of $0.090\,$m. Calculate:

  a the pressure on the plug due to the water,

  b the force needed to remove the plug from the outlet; use density of water $= 1000\,$kg/m³.

KEY POINTS

1 The pressure in a liquid increases with increase of depth and of density.

2 The pressure p due to the column of height h of liquid of density ρ is given by the equation $p = h\rho g$.

Supplement

Supplement

LEARNING OUTCOMES

- Describe and use a U-tube manometer
- Describe and use a mercury barometer
- Calculate gas pressure from the reading of a U-tube manometer
- Calculate atmospheric pressure from the reading of a mercury barometer

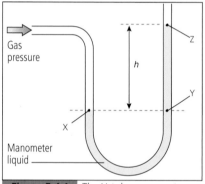

Figure 5.4.1   The U-tube manometer

STUDY TIP

You need to understand the idea of 'atmospheric pressure'.

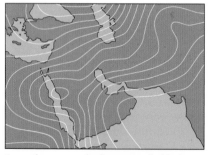

A weather map – the lines are called isobars (which join points of equal pressure)

The U-tube manometer

This instrument is used to measure the pressure of a gas, as shown in Figure 5.4.1. The pressure of the gas forces the liquid in the manometer up the open side of the U-tube until it is at rest. The height difference, h, between the two levels is a measure of the pressure of the gas.

- Because atmospheric pressure acts on the liquid on the 'open' side of the manometer:

  the gas pressure = the pressure due to the difference in liquid levels ($h\rho g$) + atmospheric pressure
- The 'excess' pressure of the gas (i.e. its pressure above atmospheric pressure) is therefore equal to the pressure due to the difference in the level of liquid on each side.

Therefore, the pressure of the gas relative to atmospheric pressure = $h\rho g$.

WORKED EXAMPLE

Use g = 10 N/kg

A U-tube manometer containing oil of density 950 kg/m³ is connected to a gas tap in a building to measure the pressure of the gas supply. When the tap is opened, the difference in the manometer levels increases to 0.26 m. Calculate the pressure of the gas supply in excess of atmospheric pressure.

Solution

Pressure due to the gas supply, $p = h\rho g$

= 0.26 m × 950 kg/m³ × 10 N/kg = 2500 Pa

Atmospheric pressure

The Earth's atmosphere extends more than 100 km above its surface. The mean pressure of the atmosphere at sea level is about 100 kPa. Atmospheric pressure varies slightly from day to day, changing with the local weather conditions as shown on the weather map. Fine clear weather is usually associated with high pressure.

The **mercury barometer** is designed to measure atmospheric pressure accurately. It consists of an inverted glass tube containing mercury with its lower end under the surface of mercury in a container open to the atmosphere. The space above the mercury in the tube is a vacuum.

The pressure of the atmosphere acting on the mercury in the container keeps the mercury in the tube. In other words, atmospheric pressure balances the pressure due to the column of mercury in the tube (Figure 5.4.2).

The height of the mercury column in the barometer is therefore a measure of the pressure of the atmosphere. For example:

- an increase of atmospheric pressure causes the mercury to rise in the tube
- a decrease of atmospheric pressure causes the mercury to fall.

The mean value of atmospheric pressure at sea level is 101 kPa. This is referred to as **standard atmospheric pressure** and it corresponds to a height of 760 mm for the mercury column in a barometer. Variations in atmospheric pressure cause the height of the mercury column to vary about the mean value by about 10–20 mm.

Supplement

By measuring the vertical height of the mercury column, atmospheric pressure can be calculated using the equation $p = h\rho g$ where ρ is the density of mercury. The height of the top of the mercury column above the mercury level in the container must be measured very accurately. Most mercury barometers are fitted with a vernier scale that measures the height to an accuracy of 0.1 mm.

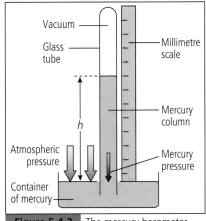

Figure 5.4.2 The mercury barometer

WORKED EXAMPLE

Use $g = 10\,\text{N/kg}$

Calculate the pressure at the base of a vertical mercury column of height 760 mm.

Density of mercury = 13 600 kg/m³.

Solution

Pressure $p = h\rho g = 0.760\,\text{m} \times 13\,600\,\text{kg/m}^3 \times 10\,\text{N/kg}$

$= 1.03 \times 10^5\,\text{Pa}$

DID YOU KNOW?

Atmospheric pressure decreases with height above sea level. The cabins of aeroplanes that fly at high altitude are pressurised so the occupants can breathe normally.

SUMMARY QUESTIONS

Use $g = 10\,\text{N/kg}$

1 a In Figure 5.4.1, atmospheric pressure acts at Z. The gas pressure to be measured acts at X. How does the pressure at X compare with the pressure at **i** Y? **ii** Z?

  b Explain how atmospheric pressure acts

   i   to keep a rubber sucker on a wall,

   ii when you drink through a straw.

Supplement

2 a Atmospheric pressure is about 100 kPa. What depth of water will give a pressure of 100 kPa?

  Use density of water = 1000 kg/m³

  b Calculate the change of atmospheric pressure, in pascals, that would cause the mercury in a barometer to rise by 1 mm.

  Use density of mercury = 13 600 kg/m³.

KEY POINTS

1 The U-tube manometer is used to measure gas pressure.

2 The mercury barometer is used to measure atmospheric pressure.

Solids, liquids and gases

LEARNING OUTCOMES

- Describe the characteristic properties of a solid, a liquid and a gas
- Describe the arrangement of particles in a solid, a liquid and a gas
- Explain the characteristic properties of a solid, a liquid and a gas

Spot the three states of matter

States of matter

Everything around us is made of matter, mostly in either a solid or a liquid or a gaseous state. Each state of matter has its own characteristic properties.

- Solids have a fixed shape and a fixed volume. In general, solid objects do not lose their shape unless they are deformed by force beyond their elastic limit.
- Liquids have a fixed volume and they flow. They change their shape to fit the shape of the container. If a liquid is stirred, it moves around internally at and below its surface. A stirred liquid gradually slows down and stops because the container 'drags' on the liquid where they are in contact.
- Gases have neither a fixed volume nor a fixed shape. Any gas can be compressed or expanded to change its volume. A gas in a container takes the shape of the container. Gases flow and they are much less dense than solids or liquids.

This table summarises the main properties of solids, liquids and gases.

	flow	shape	volume	density
solid	no	fixed	fixed	much higher than a gas
liquid	yes	fits container shape	fixed	much higher than a gas
gas	yes	fills container	can be changed	low compared with a solid or liquid

Change of state

Matter can change from one state to another. Figure 5.5.1 shows the changes between different states. The changes are brought about by heating or cooling the substance. For example:

- when liquid water in a beaker is heated sufficiently, it boils and turns to steam, which is water in the gaseous state
- when solid carbon dioxide warms up, it sublimes because it turns to vapour directly
- when liquid water freezes, it solidifies and turns to ice, which is water in the solid state
- when steam touches a cold surface, it condenses and turns to liquid water.

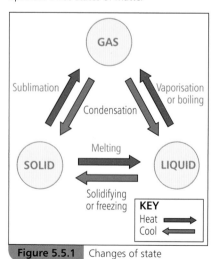

Figure 5.5.1 Changes of state

Molecules and matter

The smallest particle in a substance that can be identified as belonging to the substance is called a **molecule**. Depending on the substance, a molecule may be a single atom (in certain elements) or it may be a group of atoms. Figure 5.5.2 shows how the molecules of a substance in the solid, liquid or gaseous state are arranged.

- The molecules in a solid hold each other together in fixed positions to form a rigid structure with its own shape.
- The molecules in a liquid move about in contact with each other but they don't stay in fixed positions. This is why a liquid doesn't have its own shape and can flow.
- The molecules in a gas are far apart compared with molecules in a liquid or solid. This is why the density of a gas is much less than that of a solid or liquid. They move very rapidly throughout their container and make contact only when they collide.

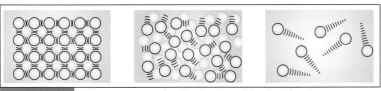

Figure 5.5.2   The arrangement of molecules in a solid, a liquid and a gas

KEY POINTS

1 Molecules in a solid are in a rigid structure.

2 Molecules in a liquid move about in contact with each other.

3 Molecules in a gas are far apart and move about at high speed.

SUMMARY QUESTIONS

1 Copy and complete the following sentences using words from the list

   gas   liquid   solid

   a A _____ has a fixed shape and volume.

   b A _____ has a fixed volume but no shape.

   c A _____ and a _____ can flow.

   d A _____ does not have a fixed volume.

2 State the scientific word for each of the following changes.

   a A mist appears on the inside of a window on a bus full of people.

   b Steam is produced from the surface of the water in a pan when the water is heated.

   c Food taken from a freezer thaws out.

   d An ice cube in a beaker of warm water gradually disappears.

PRACTICAL

Changing state

1 Place a beaker of water on a tripod and use a Bunsen burner to heat the water. Observe that evaporation takes place at the water surface before it boils. When water boils, bubbles of steam form throughout the water and rise to the surface to release the steam.

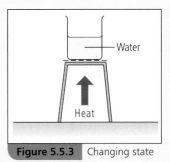

Figure 5.5.3   Changing state

2 Switch the gas burner off and hold a cold object above the beaker. Observe condensation of steam from the beaker on the object.

STUDY TIP

The next time you see ice melt or water boil, think about what is happening to the water molecules.

Supplement

LEARNING OUTCOMES

- Recognise how the temperature of a gas affects the average speed of its molecules
- Explain in terms of molecules why a gas exerts pressure
- Explain in terms of molecules, the characteristic properties of a solid, a liquid and a gas

STUDY TIP

See page 50 for internal energy.

Molecules in a gas

The pressure of a gas on a surface is due to the impacts of the gas molecules with the surface. The gas molecules move about very fast in random directions. Imagine a squash court with lots of squash balls flying about, rebounding from the walls, the floor and the ceiling. Now imagine the squash court and the balls scaled down to a thousandth of a millionth of their size. On this scale, the balls would be like gas molecules moving about at random.

The molecules in a gas collide repeatedly with each other and with the surface of their container, rebounding after each collision. They move at different speeds which can change each time they collide with each other. Each impact with the surface exerts a tiny force on the surface, as shown in Figure 5.6.1. Millions of millions of such impacts happen every second and their overall effect is to cause a steady pressure on the surface.

The temperature of a gas affects the speed of the gas molecules. If some gas in a sealed container is heated, its temperature increases because its molecules gain more kinetic energy. Therefore, the average speed of the gas molecules increases. The temperature of a gas is a measure of the average kinetic energy of the gas molecules. The molecules still move about at random but their average speed is greater. Cooling a gas lowers the average speed of the molecules.

Comparison of molecules in solids, liquids and gases

In a solid, the molecules are arranged in a three-dimensional structure.

- There are strong forces of attraction between the molecules.
- Each molecule vibrates about a mean position that is fixed.
- When a solid is heated, its internal energy increases as the molecules gain energy and vibrate more. If heated sufficiently, the solid melts (or sublimes) because the molecules have gained enough energy to break away from the structure.

Supplement

Figure 5.6.1 Gas molecules in a box

DID YOU KNOW?

'Random' means unpredictable or haphazard. The speed and direction of motion of a gas molecule is unpredictable because its collisions with other molecules cannot be predicted.

Molecular model of ice. Each water molecule consists of two hydrogen atoms (white) and one oxygen atom (red)

In a liquid, the molecules are in contact with each and they move around each other.

- There are still forces of attraction between the molecules but they are not strong enough to hold the molecules in a rigid structure. So a liquid can flow and has no fixed shape.
- The forces of attraction are strong enough to stop the molecules moving away from each other completely at the surface.
- When a liquid is heated, its internal energy increases because the molecules move about faster. Some of the molecules gain enough energy to break away from the other molecules. The molecules that escape from the liquid become a gas.

In a gas, the molecules are much further apart on average than in a solid or a liquid. This is why the density of a gas is much less than that of a liquid.

- The forces of attraction between the molecules are negligible. So a gas can flow and has no fixed shape or volume.
- The molecules move about at high speed, colliding with each other and the internal surface of their container.
- When each molecule hits the container surface, its momentum changes and it exerts a force on the container. The pressure of the gas on the container surface is due to all the molecules striking the surface.
- When a gas is heated, its internal energy increases because its molecules gain kinetic energy and move faster on average. This causes the pressure of the gas to increase because the molecules collide with the container surface with more force.

Molecules in water. Note the water molecules above the water surface that have broken away from the surface

KEY POINTS

1 Increasing the temperature of a gas increases the average speed of its molecules.

2 The pressure of a gas on a surface is caused by its molecules repeatedly hitting the surface.

SUMMARY QUESTIONS

1 Explain the following statements in terms of molecules.

  a A gas exerts a pressure on any surface it is in contact with.

  b Heating a solid makes it melt.

2 The table below lists the properties of the molecules in four different substances. State with a reason whether each substance is a solid, a liquid or a gas or does not exist.

substance	distance between the molecules	particle arrangement	movement of the molecules
A	close together	not fixed	move about
B	far apart	not fixed	move about
C	close together	fixed	vibrate
D	far apart	fixed	vibrate

Gas pressure and temperature

LEARNING OUTCOMES

- Describe how the pressure of a gas in a sealed container is affected by changing the temperature of the gas
- Explain why raising the temperature of a gas in a sealed container increases its pressure
- Describe evidence that gas molecules move about at random

STUDY TIP

The link between gas temperature and the kinetic energy of the molecules is important to know.

In the kitchen

Never heat food in a sealed can. The can is likely to explode because the pressure of gas inside it increases with an increase in temperature. Increasing the temperature of any sealed container increases the pressure of the gas inside it. This is because:

- the gas molecules move about very fast in random directions – the pressure of the gas is due to the impacts of the gas molecules with the container surface
- the heat energy supplied to the gas to raise its temperature is supplied to the gas molecules as kinetic energy.

Therefore, the average kinetic energy of the gas molecules increases when the gas is heated. As a result, the average speed of the molecules therefore increases.

When the temperature of a gas is increased, the molecules on average move about faster inside the container. They hit the surfaces harder and more often, so, if the gas is in a sealed container, its pressure increases.

PRACTICAL

Investigating how the pressure of gas varies with temperature

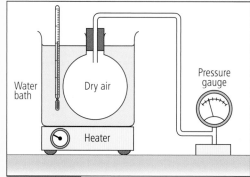

Figure 5.7.1 Measuring gas pressure at different temperatures

1 Figure 5.7.1 shows dry air in a sealed flask connected to a pressure gauge. The flask is in a large beaker of water which is heated to raise the temperature of the gas. The water is heated in stages to raise the temperature in stages. At each stage, the water is stirred to ensure its temperature is the same throughout. The temperature of the water is measured using the thermometer. The pressure is read off the pressure gauge.

2 If the measurements are plotted on a graph of pressure against temperature in °C, the results give a straight line graph as shown in Figure 5.7.2. This shows that the increase of pressure is the same for equal increases of temperature.

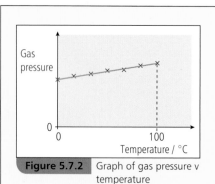

Figure 5.7.2 Graph of gas pressure v temperature

Observing random motion

Individual molecules are too small to see directly. However, we can see their direct effects by observing the motion of smoke particles in air. Figure 5.7.3 shows how we can do this using a smoke cell and a microscope. The smoke particles move about haphazardly and follow erratic (irregular and unpredictable) paths.

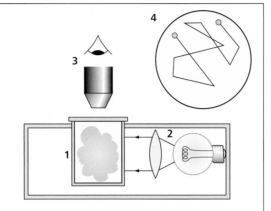

1 A small glass cell is filled with smoke

2 Light is shone through the cell

3 The smoke is viewed through a microscope

4 You see the smoke particles constantly moving and changing direction. The path taken by one smoke particle will look something like this

Figure 5.7.3 A smoke cell

This kind of motion is called **Brownian motion**. It is named after the botanist Robert Brown who first observed it in 1785. He used a microscope to observe pollen grains floating on water. He was amazed to see that the pollen grains were constantly moving about and changing direction as if they had a life of their own. Brown could not explain what he saw. Brownian motion puzzled scientists until the molecular theory of matter provided an explanation.

Figure 5.7.4 shows how the Brownian motion of smoke particles in air is caused. Air molecules repeatedly collide at random with each smoke particle. What we observe is the erratic motion of the smoke particles caused by the random impacts of gas molecules on each one.

The air molecules must be moving very fast to cause this effect because they are far too small to see and the smoke particles are massive compared with the gas molecules.

Supplement

SUMMARY QUESTIONS

1 Copy and complete the following sentences using words from the list.

   impacts   kinetic energy   pressure   temperature

   a The _____ of a gas can be increased by increasing its _____.

   b Reducing the _____ of a gas reduces the average _____ of its molecules.

   c The _____ of a gas is caused by repeated _____ of its molecules on the surface of its container.

2 a Explain why smoke particles in air move about erratically.

   b Explain why smoke particles in air move about faster if the temperature of the air is increased.

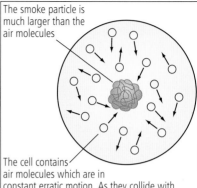

The smoke particle is much larger than the air molecules

The cell contains air molecules which are in constant erratic motion. As they collide with the smoke particle they give it a push. The direction of the push changes at random

Figure 5.7.4 Brownian motion

DID YOU KNOW?

Diffusion is the way gases and liquids spread out to occupy all the available space. Figure 5.7.5 shows what happens when a coloured gas (in this case chlorine) and air in jars spread into each other. The molecules of each gas gradually become distributed evenly throughout both jars.

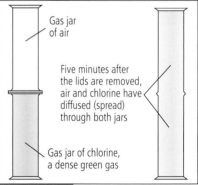

Gas jar of air

Five minutes after the lids are removed, air and chlorine have diffused (spread) through both jars

Gas jar of chlorine, a dense green gas

Figure 5.7.5 Diffusion

KEY POINTS

1 The pressure of a gas in a sealed container increases if the gas temperature is increased.

2 Brownian motion is the erratic motion of microscopic particles due to random impacts of gas molecules on each particle.

Supplement

LEARNING OUTCOMES

- Explain the difference between an unsaturated vapour and a saturated vapour
- Recognise that a liquid can be cooled by evaporation
- Explain cooling by evaporation of a liquid
- Explain the factors that can increase evaporation from a liquid

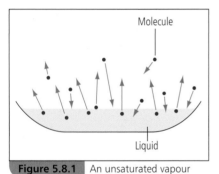

Figure 5.8.1   An unsaturated vapour

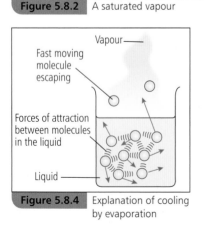

Figure 5.8.4   Explanation of cooling by evaporation

Unsaturated and saturated vapour

If a saucer of water is left in a room, the water gradually disappears. This is because water molecules move from the water in the saucer into the air space above the water. The molecules that have left the liquid spread throughout the air in the room. The water in the saucer is said to **evaporate** as it turns to water vapour. If all the water in the bowl evaporates, and the air in the room can still 'hold' more water vapour, the air is said to be **unsaturated** (Figure 5.8.1).

If a bowl containing sufficient water is left in a room that is not ventilated, evaporation occurs until the air in the room is **saturated** with water vapour. Water vapour will then condense at the same rate that water evaporates from the bowl or from condensation elsewhere. When this happens, water molecules from the water vapour in the air return to water in the liquid state at the same rate as water molecules leave the water. Figure 5.8.2 represents condensation in a room which is saturated with water vapour.

Cooling by evaporation

When a liquid evaporates, it becomes cooler. Figure 5.8.3 shows a demonstration of this effect. It can be explained in terms of molecules and their movement, as shown in Figure 5.8.4.

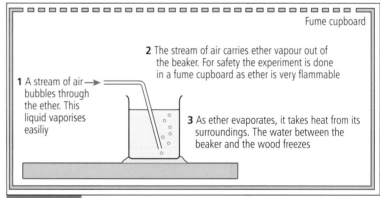

Figure 5.8.3   Cooling by evaporation

- Weak attractive forces exist between the molecules in the liquid.
- Molecules with sufficient kinetic energy break away from the attraction of the other molecules and escape from the liquid.
- After they escape the average kinetic energy of the remaining molecules in the liquid decreases so the liquid becomes cooler.
- An object in contact with an evaporating liquid is cooled by the liquid because the liquid becomes colder than the object.

Supplement

Using evaporation

1   **Salt can be obtained by allowing water to evaporate from salt water.** In hot countries, heat from the Sun causes evaporation from salt water in large, enclosed, shallow areas known as 'salt pans'. Dry salt is left in each salt pan when all the water has evaporated.

2   **An air cooling system** transfers heat energy from inside a hot building to the outside. A sealed circuit of pipes carries a volatile substance (i.e. one that easily evaporates) through the air-conditioning unit in the building to a heat exchanger outside the building and back to the air conditioning unit.

   • The liquid evaporates and cools in the pipes in the air-conditioning unit. Air blown over the pipes cools down and is circulated throughout the building.

   • The vapour in the pipes is pumped into a heat exchanger outside the building.

   • The vapour condenses in the heat exchanger and releases heat energy to the surroundings in the process.

   • The condensed vapour is then forced through valves and pipes back to the air-conditioning unit.

An air-conditioning unit

STUDY TIP

Learn the factors affecting evaporation.

Supplement

Factors affecting evaporation

The examples above show that evaporation from a liquid is increased as a result of:

• increasing the surface area and/or the temperature of the liquid,

• creating a draught of air across the liquid surface.

1   The salt pans need to be shallow so the salt water heats up easily and they need to cover a large area so evaporation readily occurs.

2   The air-conditioning system keeps the building at a constant temperature. If the building heats up, the amount of liquid evaporated in the pipes increases. As a result, more energy is transferred out of the building which cools it down.

3   A hot drink cools faster because evaporation from the liquid is increased by blowing air across it.

KEY POINTS

1   An unsaturated vapour can hold more vapour, whereas a saturated vapour cannot hold any more vapour.

2   Evaporation from a liquid occurs as a result of molecules with sufficient kinetic energy leaving the liquid.

Supplement

3   Evaporation from liquid can be increased by increasing the temperature or surface area of the liquid or by creating a draught of air across its surface.

SUMMARY QUESTIONS

1 Copy and complete the following sentences using words from the list.

   equal to   greater than   less than

   a   In an unsaturated liquid, the rate of transfer of molecules from the liquid to the vapour is _____ the rate of transfer in the opposite direction.

   b   In a saturated liquid, the rate of transfer of molecules from the liquid to the vapour is _____ the rate of transfer in the opposite direction.

   c   When the temperature of a liquid in an open container is cooled, the rate of transfer of molecules from the liquid to the vapour becomes _____ the rate of transfer in the opposite direction.

2 Explain the following statements:

   a   Wet clothes on a washing line dry out faster on a hot day than on a cold day.

   b   A person wearing wet clothes on a cold windy day is likely to feel much colder than someone wearing dry clothes.

Supplement

LEARNING OUTCOMES

- For a fixed mass of gas at constant temperature:
  - recognise how the pressure (or volume) changes affect the volume (or pressure) of the gas
  - explain in molecular terms why the pressure of a gas changes when its volume is changed
- recall and use the equation pV = constant

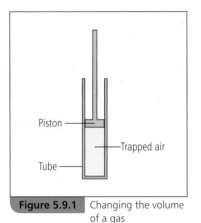

Figure 5.9.1 Changing the volume of a gas

The volume of a fixed mass of gas depends on its pressure and on its temperature. For example, suppose the air in a tube is trapped by a piston in the tube, as shown in Figure 5.9.1.

- If the piston is forced into the tube, the volume of the air in the tube decreases as the air is compressed. As a result, the pressure of the air in the tube increases. Provided the compression is carried out slowly, the temperature of the air in the tube does not change.
- If the tube is heated, for example by placing it in hot water, the air expands and forces the piston towards the open end of the tube. In this situation, the volume of the air in the tube increases. The air pressure in the tube stays the same as the air expands.

Investigating the pressure and volume of a fixed mass of air at constant temperature

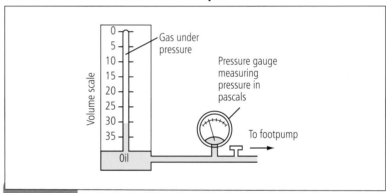

Figure 5.9.2 Testing the variation of pressure and volume of a fixed mass of air

Figure 5.9.2 shows a gas trapped by oil in an inverted glass tube. The pressure of the air in the tube is measured using a pressure gauge. The volume of the air is measured in cubic centimetres using the vertical scale alongside the tube. The pressure of the gas is increased using a foot pump. The tube has a thick wall so that it can withstand the high internal pressure of the gas.

The volume of the gas in the tube is measured at different pressures as the pressure is increased slowly from atmospheric pressure in stages. At each stage, the tap is closed so the gas in the tube does not leak out and the pressure and volume of the gas are measured when they have stopped changing. The measurements are recorded in a table as shown.

They show that:

- the pressure increases as the volume decreases,
- the pressure decreases as the volume increases.

pressure / Pa	100	120	150	180
volume / cm³	36	30	24	20
pressure × volume / Pa cm³	3600	3600	3600	3600

Explanation of the variation of pressure with volume

For a fixed mass of gas, the number of gas molecules is constant. If the temperature is constant, the average speed of the molecules is constant.

If the volume of a fixed mass of gas at constant temperature is reduced, the gas pressure increases because:

- the space the molecules move in is smaller so they do not travel as far between successive impacts on the surface,
- the molecules hit the surfaces more often so the average force of impact increases.

STUDY TIP

This is an important explanation. Try to understand it; do not just learn the words.

The measurements in the table on the previous page show that the product, pressure × volume, is constant. This is known as Boyle's law after Robert Boyle who first discovered the law in 1662.

For a fixed mass of gas at constant temperature:

pressure × volume = constant

or

pV = constant where p = the gas pressure and V = the gas volume.

Note: Rearranging the above equation gives $p = \dfrac{\text{constant}}{V}$, which means that the pressure varies in inverse proportion to the volume of the gas.

WORKED EXAMPLE

In a chemistry experiment, $0.000\,20\,m^3$ ($= 200\,cm^3$) of gas was collected in a flask at a pressure of $125\,kPa$. Calculate the volume of this mass of gas at a pressure of $100\,kPa$ and the same temperature.

Solution

Let $p_1 = 125\,kPa = 125\,000\,Pa$ and $V_1 = 0.000\,20\,m^3$, $p_1 V_1 = 125\,000\,Pa \times 0.00020\,m^3 = 25\,Pa\,m^3$.

Let $p_2 = 100\,kPa = 100\,000\,Pa$, where V_2 is the volume to be calculated.

Applying $p_2 V_2 = p_1 V_1$ therefore gives $100\,000\,Pa \times V_2 = 25\,Pa\,m^3$

Hence $V_2 = \dfrac{25\,Pa\,m^3}{100\,000\,Pa} = 0.000\,25\,m^3$ ($= 250\,cm^3$)

SUMMARY QUESTIONS

1 Copy and complete the following sentences using words from the list.

   decreases   increases   stays the same

   a When a fixed mass of gas expands at constant temperature, its volume _____ and its pressure _____.

   b When a fixed mass of gas is compressed, its pressure _____, its volume _____ and its density _____.

   c When the volume of a gas is reduced at constant temperature, the average kinetic energy of its molecules _____ and the number of impacts per second of gas molecules on the surfaces in contact with the gas _____.

2 Calculate the unknown quantity in each of the following changes involving a fixed mass of gas at constant pressure.

	initial pressure / Pa	initial volume / m³	final pressure / Pa	final volume /m³
a	100 000	0.000 20	50 000	?
b	100 000	0.000 30	?	0.000 15
c	120 000	?	100 000	0.000 60
d	?	0.000 15	60 000	0.000 45

KEY POINTS

For a fixed mass of gas at constant temperature, if its volume is decreased:

- its molecules hit the surfaces of its container more often,
- its pressure increases,
- its pressure × its volume is constant.

Supplement

85

1 Describe an experiment to study the relationship between the pressure and volume of a fixed mass of gas at constant temperature.

Include a labelled diagram and suggest any precautions that you would take to make the experiment as reliable as possible.

2 Explain the reason for the following.
  (a) A tractor has large, wide rear wheels.
  (b) A sharp knife cuts more easily than a blunt knife.
  (c) A small stone trapped in your shoe, under your foot, can be very painful.

3 The diagram shows gas molecules in a box.

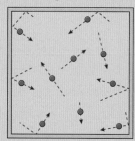

  (a) Describe the motion of the gas particles.
  (b) Comment on the forces between the particles.
  (c) The size of the container is fixed. Explain why increasing the temperature of the gas will increase its pressure.

4 (a) When a liquid evaporates quickly there is a noticeable cooling effect. Explain why this cooling occurs.
  (b) State and explain the factors that affect the rate of evaporation of a liquid.

5 In a hydraulic machine a force of 40 N is applied to a piston of area 0.40 m². The area of the other piston is 4.0 m².

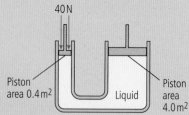

40 N

Piston area 0.4 m²   Liquid   Piston area 4.0 m²

  (a) Calculate the pressure transmitted through the liquid.
  (b) Calculate the force on the other piston.

Q1 to 4 *(Paper 1/2)*

1 A unit of pressure is
  A N   **B** N/m
  C Nm²   **D** N/m²

2 The equation used to calculate the pressure that an object exerts on the ground is
  A $P = F + A$ **B** $P = A/F$
  C $P = F/A$ **D** $P = F \times A$

3 An object with weight 6000 N has the dimensions shown in the diagram.

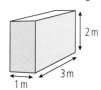

2 m
3 m
1 m

The pressure, in Pa, exerted on the ground by the object is
  A 1000   **B** 2000
  C 9000   **D** 36 000

4 Which of the following is NOT a correct statement?
  A The pressure in a liquid increases with depth
  B The pressure in a liquid is the same in all directions
  C The pressure in a liquid depends on the density of the liquid
  D The pressure in a liquid is greatest near the walls of the container

5 The relationship between the pressure and volume of a gas is given by the equation pV = constant. This applies to
  A a fixed mass of gas at any temperature
  B a fixed mass of gas at constant temperature
  C any mass of gas at any temperature
  D any mass of gas at constant temperature

(Paper 2)

6 (a) The diagram shows the apparatus used for a liquid pressure demonstration.

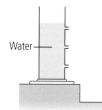

Water

Supplement

(i) Copy this diagram and draw in a line from each spout to show the flow of water that you would expect.

(ii) Write down a conclusion from the result you have drawn. *[3]*

(b) The diagram shows a similar experiment.

Write down what the result shown indicates about the pressure in a liquid. *[2]*

(c)

The diagram shows a cross-section through a dam wall. Explain briefly why the dam wall is thicker at the bottom than the top.
[1]

(Paper 3)

7 (a) Copy and complete the table to show the properties of solids, liquids and gases.

	can be poured (yes or no)	shape
solid	no	has its own fixed shape
liquid		
gas		

[4]

(b) Copy and complete the following sentences by filling in the spaces.
  (i) When water is heated to a temperature of _____ it _____ .
  (ii) When solid carbon dioxide is surrounded by air at room temperature it _____ and turns directly to _____ .
  (iii) When water _____ it solidifies and turns to _____ . *[5]*

(Paper 3)

8 In an experiment to observe Brownian motion, smoke particles in a small glass container are illuminated and observed through a microscope. The diagram shows one such smoke particle.

(a) Draw a diagram to show the way in which you would expect to see the smoke particle move.

(b) Describe in a few words the motion of the smoke particle.

(c) Explain how the smoke particle is made to move in this way. *[5]*

(Paper 4)

9 The diagram shows a dish of water at room temperature and the motion of a few of the water molecules.

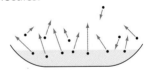

(a) (i) Explain briefly the process of evaporation pictured.

(ii) Suggest three factors that will increase the rate of evaporation. *[5]*

(b) If some gas in a sealed container of constant volume is heated, the pressure of the gas increases. Explain, in terms of the molecules, why the increase in temperature causes the pressure to increase. *[3]*

(c) In an experiment to study the pressure and volume of a fixed mass of gas at constant temperature, the starting volume is 30 cm³ and the pressure is 200 kPa. The pressure is reduced to 120 kPa. Calculate the volume at this pressure. *[3]*

(Paper 4)

Thermal expansion

Supplement

LEARNING OUTCOMES

- Recognise that solids, liquids and gases expand when heated
- Describe and explain some applications of thermal expansion
- Describe and explain some consequences of thermal expansion
- Appreciate that gases expand much more than solids and liquids

Filling a hot-air balloon

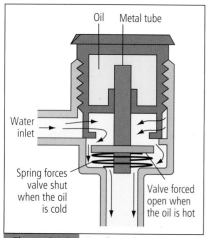

Figure 6.1.2 A radiator thermostat

Comparing the thermal expansion of liquids and gases

Most substances expand when heated. If air did not expand when heated, a hot-air balloon would never fill up and take off. A burner under the balloon causes the balloon to fill with hot air which then lifts the balloon. The expansion of a substance due to increasing its temperature is called **thermal expansion.**

A simple example of the thermal expansion of air in a test tube is shown in Figure 6.1.1. If an identical test tube is filled completely with water, as shown in Figure 6.1.1, we can compare the thermal expansion of air with that of water. The air in the air-filled test tube is trapped by a 'thread' of water in the narrow tube. In each case, expansion of the air or water in the test tube causes the level of the water in the narrow tube to move towards the open end of the narrow tube.

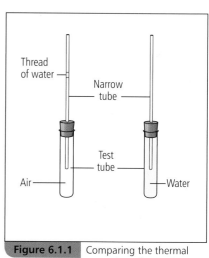

Figure 6.1.1 Comparing the thermal expansion of air and water

To ensure both test tubes are heated exactly the same, both test tubes could be placed next to each other in strong sunlight (or in the same beaker of warm water). The results show that air expands much more than water for the same temperature rise. The same result applies to a comparison between any gas and any liquid.

A radiator thermostat in a car makes use of the expansion of oil when it becomes warm. Figure 6.1.2 shows how it works. The valve stays closed until it has been warmed up by thermal energy from the engine. The metal tube is forced out of the oil chamber as the oil becomes hotter and expands. The movement of the metal tube opens the valve, allowing the hot water to flow to the radiator.

Thermal expansion of solids

A jar with a tight-fitting lid is sometimes difficult to open. Warming the lid in warm water may help to open the jar. The increase of temperature of the lid causes it to expand just enough to help you unscrew the lid.

Expansion gaps are necessary in buildings, bridges and railway tracks to allow for thermal expansion. Outdoor temperatures can change by as much as 50 °C between summer and winter.

A 100 m concrete bridge span would expand by about 5 cm if its temperature increased by 50 °C. Without expansion gaps, the bridge would buckle. The gaps are usually filled with soft material such as rubber to prevent rubble falling in.

Steel tyres are fitted on train wheels by heating the tyre so it expands, then fitting it on the wheel. As the tyre cools, it contracts so it fits very tightly on the wheel.

Bimetallic strips are used in thermostats in devices fitted with thermal cut-out switches or valves. A bimetallic switch consists of a strip of two different metals such as brass and steel stuck together. When the temperature of the strip rises, one metal expands more than the other (e.g. brass expands more than steel) so the strip bends. This can be used to switch on or off an electrical device. Figure 6.1.3 shows a bimetallic strip in a fire alarm. When heated, the strip bends towards the contact screw and when it touches it, the circuit is completed and the bell rings.

In a heater thermostat, when the strip becomes hot, it bends away from the contact screw and loses contact with it. As a result, the heater is switched off.

Supplement

In general, the volume of a solid increases by no more than about 0.01% for a temperature increase of 1°C. Most liquids expand slightly more.

Gases at constant pressure expand about 30 times more. This is because the molecules in a solid or liquid are unable to break free from each other, as they have much less kinetic energy than in a gas.

Expansion gaps

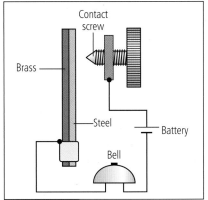
Figure 6.1.3 A bimetallic strip in an alarm circuit

SUMMARY QUESTIONS

1 Figure 6.1.4 shows a thermostat in a gas oven. When the oven overheats, the brass tube expands more than the invar rod. Explain why this change reduces the flow of gas through the valve.

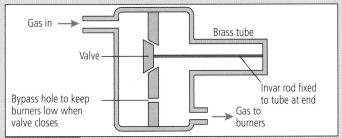
Figure 6.1.4

2 a Explain why expansion gaps are **i** necessary between concrete sections in buildings, **ii** usually filled with soft material such as rubber.

b i Explain why a bimetallic strip bends when it is heated.

ii A bimetallic thermostat is used to switch off the electricity supply to a heater if the heater overheats. With the aid of a diagram, describe such a thermostat and explain how it works.

STUDY TIP

There are many examples of uses for expansion and precautions taken to prevent damage due to expansion. The important thing is to understand the principles.

KEY POINTS

1 Applications of thermal expansion include liquid-in-glass thermometers and thermostats.

2 Thermal expansion must be allowed for in buildings and bridges.

3 Gases expand much more than solids and liquids.

Supplement

Thermometers

Supplement

LEARNING OUTCOMES

- Describe the fixed points for the Celsius scale of temperature
- Describe a liquid-in-glass thermometer and how it works
- Explain how a thermocouple thermometer works and its uses

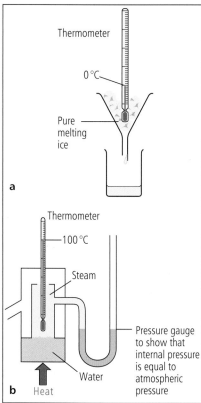

Figure 6.2.1 Calibrating a liquid-in-glass thermometer: (a) ice point, (b) steam point

STUDY TIP

Make sure you understand the difference between heat and temperature.

Temperature

Temperature is a measure of hotness. When a weather forecast tells you that tomorrow's outdoor temperature could be 10 °C lower than today, you can expect a much cooler day. **Fixed points** are used to define a scale of temperature. These points are 'degrees of hotness' that can be reproduced precisely. They are usually melting points or boiling points of pure substances.

The Celsius scale of temperature, denoted by °C, is defined by two fixed points which are:

- Ice point at 0 °C, the temperature at which pure ice melts, and
- Steam point at 100 °C, the temperature at which pure water boils at standard atmospheric pressure.

Figure 6.2.1 shows how a liquid-in-glass thermometer is calibrated (i.e. marked) at ice point and at steam point.

The length between the two calibrations can then be marked with 100 equal intervals, each corresponding to a temperature change of 1 degree Celsius. The temperature according to the thermometer is then read from the position of the liquid meniscus in the narrow section of the thermometer.

Types of thermometer

Every thermometer makes use of a physical property that varies with temperature. This property is referred to as the **thermometric property** of the thermometer. For example, the thermometric property of a liquid-in-glass thermometer is the thermal expansion of the liquid.

The liquid-in-glass thermometer consists of a thin glass bulb joined to a capillary tube with a narrow bore which is sealed at its other end. The liquid fills the bulb and the adjoining section of the capillary tube. When the bulb becomes warmer:

- the liquid in it expands more than the bulb so some of the liquid in the bulb is forced into the capillary tube
- the thread of liquid in the capillary tube increases in length
- the thinner the bulb wall is, the faster the response of the thermometer will be when the temperature changes.

The liquid used usually contains mercury or coloured alcohol. Alcohol has a lower freezing point than mercury so it is more suitable for low-temperature measurements.

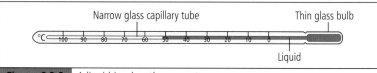

Figure 6.2.2 A liquid-in-glass thermometer

Thermocouple thermometers are electrical thermometers which make use of the voltage that develops when two different metals are in contact. This voltage varies with temperature. An iron wire and two copper wires may be used to make a thermocouple thermometer, as shown in Figure 6.2.3. One of the junctions is maintained at 0 °C and the other junction is used as the temperature probe. The voltmeter (which can be analogue, as shown, or digital) can be calibrated directly in °C.

Because of the small size of a thermocouple junction, thermocouple thermometers are used to measure rapidly changing temperatures. In addition, they can be used to measure much higher temperatures than liquid-in-glass thermometers. Also, the voltage of a thermocouple can be measured and recorded automatically.

The table below compares the liquid-in-glass thermometer and a thermocouple thermometer.

type of thermometer	thermometric property	examples of use
liquid-in-glass	thermal expansion of a liquid	home, office, greenhouse, hospital (clinical thermometer)
thermocouple	voltage between two different metals in contact	food heated in an oven, temperature variation of gas flow, data logging, remote locations

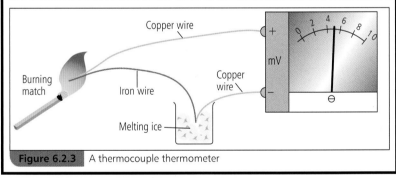

Figure 6.2.3   A thermocouple thermometer

DID YOU KNOW?

Body temperature for a healthy person is about 37 °C. Clinical thermometers are designed to measure body temperature so they are graduated in intervals of 0.1 °C from about 32 °C to 42 °C. A liquid-in-glass clinical thermometer has a constriction (i.e. a very narrow section) near the bulb to prevent the liquid returning to the bulb after the thermometer is removed from the patient.

STUDY TIP

Do you know and understand the advantages of using thermocouple thermometers?

KEY POINTS

1 The temperature of an object is a measure of its degree of hotness.

2 The fixed points of the Celsius scale are ice point (0 °C) and steam point (100 °C).

3 Each type of thermometer depends on a physical property that varies with temperature.

SUMMARY QUESTIONS

1 Copy and complete the sentences below using words from the following list.

   expansion   temperature   voltage

   a   The degree of hotness of an object is a measure of its _____.

   b   The liquid-in-glass thermometer makes use of the _____ of the liquid when its _____ changes.

   c   In a thermocouple thermometer, a change of _____ causes its _____ to change.

2 State and explain which one of the following types of thermometer is most suitable for the temperature measurements **a–c** below.

   liquid-in-glass   thermocouple

   a   the temperature change of exhaled air

   b   the temperature inside cooked meat in an oven

   c   the temperature in a greenhouse.

More about thermometers

Supplement

Supplement

LEARNING OUTCOMES

- Describe the thermal expansion of a gas at constant pressure

- Explain what is meant by the range, sensitivity and linearity of a thermometer

Thermal expansion of a gas

A gas at constant pressure expands when it is heated. We can use this property to measure temperature. Figure 6.3.1 shows how we can measure the volume of gas at constant pressure at different temperatures. The gas is air trapped in the tube by a liquid thread. As the water bath is warmed, the gas trapped in the glass tube becomes hotter and expands, pushing the liquid thread up the tube. The length of the gas column increases. This is a measure of the gas volume.

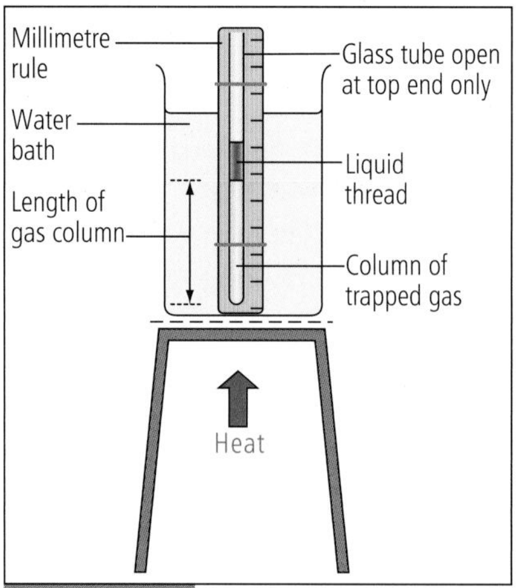

Figure 6.3.1 Measuring the expansion of a gas using a water bath

The gas thermometer

We can use the arrangement in Figure 6.3.1 as a simple gas thermometer. If the length of the trapped gas column is measured at 0 °C and 100 °C, we can calibrate the thermometer by dividing the increase in length into 100 equal intervals, as explained on page 90.

The gas thermometer is used as a standard thermometer to calibrate other thermometers between the fixed points. This is because gas thermometers with different gases will always agree with each other between the fixed points. This doesn't happen with other types of thermometer because:

- in a liquid-in-glass thermometer, the thermal expansion of a liquid depends on the choice of the liquid
- in a thermocouple thermometer, the voltage of a thermocouple depends on the two metals used.

When the scale of a thermometer is calibrated, the interval between the 0 °C and the 100 °C reading is usually divided into 100 equal 'degrees'. Suppose this is done for a gas thermometer and a liquid-in-glass thermometer.

The graph in Figure 6.3.2 shows how the temperature of the liquid-in-glass thermometer compares with the temperature of the gas thermometer. Between 0 °C and 100 °C, the liquid-in-glass thermometer gives a lower reading. This is because the thermal expansion of the liquid per 1 °C rise of temperature is not constant. In other words, a graph of the thermal expansion of a liquid against temperature in °C is non-linear (i.e. not a straight line). This is shown by the red line in Figure 6.3.2.

Choosing a thermometer

Which thermometer to use in a given situation depends on various factors including the following items.

1   **The range of temperatures** to be measured; this is the temperature range to be measured from the lowest to the highest temperature. Alcohol freezes at −114 °C and boils at 78 °C, whereas mercury freezes at −38 °C and boils at 357 °C. An alcohol

STUDY TIP

'Range', 'sensitivity' and 'linearity' are technical terms. You must learn what each one means and not confuse them.

thermometer is therefore better than a mercury thermometer for low temperature measurements but not for high temperatures.

2 **The sensitivity** of each available thermometer. The sensitivity of a thermometer is the extent of change in the thermometric property for a 1 °C rise of temperature. For example, alcohol expands about five times more than mercury for the same change of temperature. Therefore, an alcohol thermometer is much more sensitive than a mercury thermometer of the same dimensions. A thermometer with a thin capillary tube is more sensitive than one with the same liquid that has a wider capillary tube.

3 **The linearity** of each type of thermometer. For a linear scale, the thermometric property would need to change by equal amounts at equal intervals along the scale. As shown in Fig 6.3.2, a liquid-in-glass thermometer and a thermocouple thermometer give readings that differ from the corresponding readings of a standard thermometer (e.g. a gas thermometer). The linearity of the scale is the extent of the difference. The bigger the difference, the poorer the linearity of the scale.

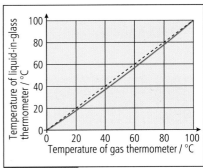

Figure 6.3.2 Comparison of a liquid-in-glass thermometer and a gas thermometer

The **accuracy** of a thermometer depends on its sensitivity and its linearity. If the non-linearity of a thermometer is small, a 'calibration' graph such as Figure 6.3.2 may be used to find the temperature in °C. A very sensitive thermometer would give inaccurate readings if non-linearity is not taken into account.

Scientific liquid-in-glass thermometers are usually graduated to match the thermal expansion of the liquid. The graduation marks are spaced unevenly along the scale so as to give readings directly in °C.

SUMMARY QUESTIONS

1 In the investigation shown in Figure 6.3.1, explain why:
  a the length of the column of trapped air is a measure of the volume of the gas
  b the water bath must be stirred before its temperature is measured
  c the thermometer should not be removed from the water when the water temperature is being measured
  d the trapped air stays at constant pressure when it is heated.

2 a A liquid-in-glass clinical thermometer is marked with a scale from 32 °C to 42 °C which covers a distance of 80 mm. A liquid-in-glass laboratory thermometer is marked with a scale from 0 °C to 100 °C, which covers a distance of 160 mm. State and explain which thermometer:
  i has the greater range,
  ii is more sensitive.
  b The following measurements were made when the voltage from a thermocouple thermometer was measured at different temperatures.

voltage / millivolts	0	1.40	2.85	4.35	5.90	7.50
temperature / °C	0	20	40	60	80	100

  i Plot a graph of the voltage on the y-axis against temperature in °C on the x-axis.
  ii Calculate the voltage midway between ice point and steam point and use the graph to determine the temperature at this voltage.

KEY POINTS

1 The volume of a gas at constant pressure increases when its temperature is increased.

2 The range of a thermometer is from the lowest to the highest temperature it can measure.

3 The sensitivity of a thermometer is the extent of change in its thermometric property for a 1°C change of temperature.

4 The greater the linearity of a thermometer, the closer its readings agree with a standard thermometer.

- Explain what is meant by thermal capacity
- Recall and use $E = mc\,\Delta T$ in specific heat calculations
- Describe an experiment to measure the specific heat capacity of a body

STUDY TIP

See page 50 for internal energy.

Thermal capacity

A piece of metal in strong sunlight can become very hot. A stone of equal mass would not become as hot even though it gains the same amount of internal energy. The stone needs more energy to give it the same temperature rise as the piece of metal. It has a greater thermal capacity than the piece of metal.

The thermal capacity of an object is the energy that must be supplied to it to raise its temperature by 1 °C.

If an object is supplied with energy E and its temperature increases by ΔT

its thermal capacity $= \dfrac{E}{\Delta T}$

For example, if an object is supplied with 5000 J of energy and its temperature increases by 2 °C, its thermal capacity is 2500 J/°C ($= 5000\,\text{J}/2\,°\text{C}$).

PRACTICAL

Investigating heating

1 Use a low voltage electric heater connected to a power supply and a joulemeter to heat some water in a plastic cup, as shown in Figure 6.4.1. Measure the temperature rise of the water using a thermometer and measure the energy supplied using the joulemeter.

2 Repeat the test using twice as much water. Typical results are shown in the table below. They show that the larger the mass of water:

- the smaller the temperature rise,
- the greater its thermal capacity.

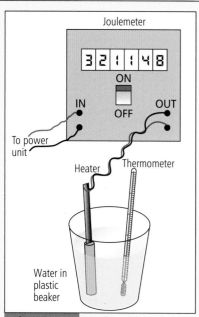

Figure 6.4.1 Heating water

mass of water (in kg)	0.10	0.20
energy supplied (in J)	6000	6000
temperature rise (in °C)	12	6
thermal capacity (J/°C)	500	1000

Specific heat capacity

Supplement

When a substance is heated, its temperature rise depends on:

1 **the amount of energy supplied**
2 **the mass of the substance**
3 **the nature of the substance.**

The specific heat capacity, c, of a substance is defined as the energy needed to raise the temperature of 1 kg of the substance by 1 °C.

The unit of specific heat capacity is the joule per kilogram per °C or J/(kg °C) in symbols.

If mass m of a substance is supplied with energy E, its specific heat capacity, $c = \dfrac{E}{m(\Delta T)}$ where ΔT is the temperature increase.

Rearranging this equation gives: energy supplied $\boldsymbol{E = mc(\Delta T)}$

Note from the above equation, thermal capacity $= \dfrac{E}{\Delta T} = mc$

Measuring the specific heat capacity of a metal

1   A block of the metal of known mass m in an insulated container is used. A 12 V electrical heater is used to heat the metal by supplying a measured amount of electrical energy as shown in Figure 6.4.2. The initial temperature T_1 of the block is measured using the thermometer.

2   A joulemeter may be used as shown in Figure 6.4.1 to measure the energy supplied to the circuit. The joulemeter reading is set to zero and the power supply to the heater is then switched on for a certain time. After this time, the power supply is switched off and the joulemeter reading and the final temperature T_2 of the block are measured.

3   The specific heat capacity of the metal can then be calculated using the equation:

specific heat capacity: $c = \dfrac{E}{m(\Delta T)}$, where $\Delta T = T_2 - T_1$

Note An ammeter, a voltmeter and a stopwatch may be used instead of the joulemeter to measure the electrical energy supplied to the block. The circuit for this is shown opposite. The ammeter is used to measure the heater current and the voltmeter is used to measure the heater voltage.

As explained in Topic 12.5, the electrical energy supplied E = heater current $I \times$ heater p.d. $V \times$ heating time t

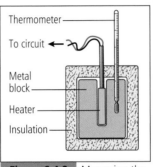

Figure 6.4.2   Measuring the specific heat capacity of a metal block

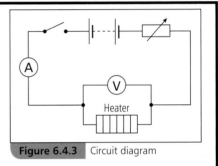

Figure 6.4.3   Circuit diagram

KEY POINTS

1   The thermal capacity of an object is the energy that must be supplied to it to raise its temperature by 1 °C.

2   The specific heat capacity, c, of a substance is defined as the energy needed to raise the temperature of 1 kg of the substance by 1 °C.

3   To raise the temperature of mass m of a substance by ΔT, the energy needed, $E = mc\,\Delta T$.

SUMMARY QUESTIONS

1   A bottle of water and a large bucket of water are left in strong sunlight.

   a   Which one warms up faster? Give a reason for your answer.

   b   Which one has the larger thermal capacity? Explain your answer.

2   Use the information below to answer this question.

substance	water	aluminium	lead	concrete
specific heat capacity /J/(kg °C)	4200	900	130	850

   a   Explain why a mass of lead heats up quicker than an equal mass of aluminium.

   b   Calculate the energy needed:

     i   to raise the temperature of 0.20 kg of aluminium from 15 °C to 40 °C

     ii   to raise the temperature of 0.40 kg of water from 15 °C to 40 °C

     iii   to raise the temperature of 0.40 kg of water in a 0.20 kg aluminium pan from 15 °C to 40 °C.

   c   A storage heater consists of a 20 kg concrete storage block in an aluminium case of mass 4.0 kg. Calculate the energy needed to heat a storage heater from 10 °C to 45 °C.

Change of state

- Describe what is meant by melting point and boiling point
- Recognise that energy is needed to melt a solid or boil a liquid
- Explain the difference between boiling and evaporation

Melting points and boiling points

When pure ice is heated and it melts, its temperature remains at 0 °C until all the ice has melted. When water is heated and it boils at atmospheric pressure, its temperature remains at 100 °C.

For any pure substance undergoing a change of state, its temperature stays the same. As shown in the table below, we refer to this temperature as the melting point or the boiling point of the substance, according to the type of change. The temperature at which a liquid changes to a solid is referred to as its freezing point and is the same temperature as the melting point of the solid.

STUDY TIP

Students are often confused about the difference between evaporation and boiling.

change	initial and final state	temperature
melting	solid to liquid	melting point
freezing (also called solidification)	liquid to solid	melting point
boiling	liquid to vapour	boiling point
condensation	vapour to liquid	boiling point

Measurement of the melting point of a substance

1 The substance in its solid state is placed in a test tube in a beaker of water, as shown in Figure 6.5.1a.

2 The water is heated and the temperature of the substance is measured when it melts.

3 If its temperature is measured every minute, the measurements can be plotted as shown in Figure 6.5.1b. The melting point is the temperature of the flat section of the graph, because this is when the temperature remains constant as the substance melts.

4 The same arrangement without the beaker of water may be used to find the boiling point of a liquid.

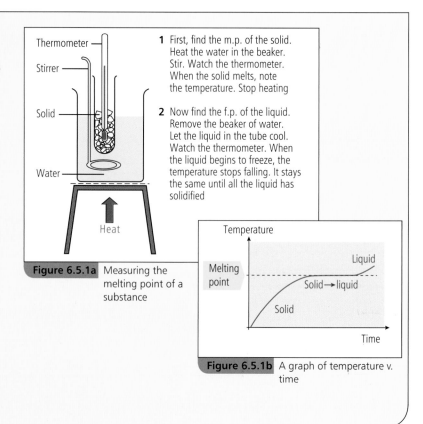

1 First, find the m.p. of the solid. Heat the water in the beaker. Stir. Watch the thermometer. When the solid melts, note the temperature. Stop heating

2 Now find the f.p. of the liquid. Remove the beaker of water. Let the liquid in the tube cool. Watch the thermometer. When the liquid begins to freeze, the temperature stops falling. It stays the same until all the liquid has solidified

Figure 6.5.1a Measuring the melting point of a substance

Figure 6.5.1b A graph of temperature v. time

Energy and change of state

Suppose a beaker of ice below 0 °C is heated steadily so that the ice melts and then the water formed from the melted ice boils. Figure 6.5.2 shows how the temperature changes with time.
The temperature:

1 increases until it reaches 0 °C when the ice starts to melt at 0 °C, then

2 remains constant at 0 °C until all the ice has melted, then

3 increases from 0 to 100 °C until the water in the beaker starts to boil at 100 °C, then

4 remains constant at 100 °C as the water turns to steam.

STUDY TIP

The shape of the graph in 6.5.1b is unusual. You must be able to describe the significance of the different regions.

The energy supplied to a substance when it changes its state is called **latent heat**. 'Latent' means 'hidden'. The energy supplied to melt or boil it is 'hidden' by the substance because its temperature does not change at the melting point or the boiling point.

Most pure substances produce a graph with similar features to Figure 6.5.2. Note that:

1 **fusion** is often used to describe melting as different solids can be fused together when they melt

2 **evaporation** from a liquid occurs at its surface when the liquid is heated below its boiling point. At its boiling point, a liquid boils because bubbles of vapour form inside the liquid and rise to the surface to release the gas.

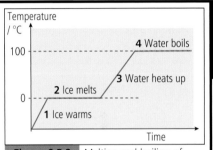

Figure 6.5.2 Melting and boiling of water

STUDY TIP

Boiling occurs only at the boiling point. Evaporation takes place at any temperature below this.

SUMMARY QUESTIONS

1 Copy and complete the sentences **a** to **d** below using words from the following list.

decreases   increases   stays the same

a When a liquid boils, its temperature _____.

b When a solid is cooled below its melting point, its temperature _____.

c When a liquid condenses, its temperature _____.

d When a solid is heated at its melting point, its temperature _____.

2 A pure solid substance X was heated in a tube and its temperature was measured every 30 s. The measurements are given in the table below.

time / s	0	30	60	90	120	150	180	210	240	270	300
temp / °C	20	35	49	61	71	79	79	79	79	86	92

a i Use the measurements in the table to plot a graph of temperature on the y-axis against time on the x-axis.

  ii Use your graph to find the melting point of X.

b Describe the physical state of the substance as it was heated from 60 °C to 90 °C.

KEY POINTS

1 For a pure substance:

  • its melting point is the temperature at which it melts or solidifies

  • its boiling point is the temperature at which it boils or condenses.

2 Energy is needed to melt a solid or boil a liquid.

3 Boiling occurs throughout a liquid at its boiling point. Evaporation occurs from the surface of a liquid when its temperature is below the boiling point.

LEARNING OUTCOMES

- Explain what is meant by latent heat of fusion and latent heat of vaporisation
- State what is meant by specific latent heat of fusion and of vaporisation in latent heat calculations
- Recall and use the equation $E = ml$

Latent heat of fusion

When a solid substance is heated

- **below its melting point**, its temperature increases until the melting point is reached. Before it reaches this point, the energy supplied to the solid increases the kinetic energy of its molecules. As a result, the molecules vibrate more and more.
- **at its melting point**, it melts and turns to liquid. Its temperature remains constant until it has all melted. The energy supplied is referred to as **latent heat of fusion**. It is the energy used by the molecules to break free from each other. Their potential energy increases so the internal energy of the substance increases even though their kinetic energy is unchanged (as the temperature does not change).

If a substance in its liquid state is cooled, it will solidify at its melting point. When this happens, the molecules bond together into a rigid structure and they release latent heat in the process.

The **specific latent heat of fusion**, *l*, of a substance is the energy needed to melt 1 kg of the substance at its melting point. The unit of specific latent heat of fusion is the joule per kilogram (J/kg).

PRACTICAL

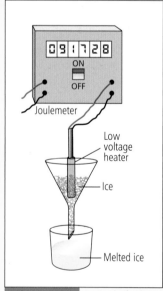

Figure 6.6.1 Instead of using a joulemeter, the energy supplied to the heater may also be measured using the circuit and information in Topic 6.4, Figure 6.4.3.

Measurement of the specific latent heat of ice

In this experiment, a low voltage heater is to be used to melt crushed ice in a funnel. The melted ice is collected using a beaker under the funnel, as shown in Figure 6.6.1. A joulemeter is used as shown to measure the energy supplied to the heater.

To take account of heat transfer from the surroundings, the mass of ice melted in a certain time must be measured with the heater off then with it on. The difference in the two measurements gives the mass of ice melted due to the heater only.

1. With the heater off, water from the funnel is collected in the beaker for a measured time (e.g. 10 min). The mass of the beaker and water, m_1, is then measured. The beaker is then emptied for the next stage.

2. With the heater on, the procedure is repeated for exactly the same time. The joulemeter readings before and after the heater is switched on are recorded. After the heater is switched off, the mass of the beaker and the water, m_2, is measured once more.

To calculate the specific latent heat of fusion of ice, note that:

- the mass of ice melted due to the heater, $m = m_2 - m_1$
- the energy supplied E to the heater = the difference between the joulemeter readings
- the specific latent heat of fusion of ice, $l = \dfrac{E}{m} = \dfrac{E}{m_2 - m_1}$

Note that rearranging this equation gives $E = ml$

Latent heat of vaporisation

When a liquid is heated:

- **above its freezing point,** its temperature increases until the boiling point is reached. Before it reaches this point, the energy supplied to the liquid increases the kinetic energy of its molecules. As a result, the molecules move about faster and faster.

- **at its boiling point,** it turns to vapour. Its temperature remains constant until it has all melted. The energy supplied is referred to as **latent heat of vaporisation**. It is the energy used by the molecules to break away from the liquid. As explained opposite, their potential energy increases without change of their kinetic energy so the internal energy of the substance increases.

If a substance in its vapour state is cooled, it will condense to become liquid at the same temperature as its boiling point. When this happens, the molecules bond together loosely and latent heat is released.

The specific latent heat of vaporisation, l, of a substance is the energy needed to change 1 kg of the substance at its boiling point from liquid to vapour. The unit of specific latent heat of vaporisation is the joule per kilogram (J/kg).

If energy E is supplied to a solid at its melting point or to a liquid at its boiling point and mass m of the substance changes its state without change of temperature,

its specific latent heat of fusion or vaporisation, $l = \dfrac{E}{m}$

PRACTICAL

Measurement of the specific latent heat of steam

In this experiment, the arrangement shown in Figure 6.6.2 can be used. The low voltage heater is switched on to bring the water to its boiling point. When the water is boiling, the joulemeter reading and the top pan balance reading are measured and then re-measured after a set time (e.g. 5 min).

The energy supplied, E, during this time = the difference between the joulemeter readings.

The mass of water boiled away in this time, m = the difference between the readings of the top pan balance.

Calculate steam's specific latent heat with the equation $l = \dfrac{E}{m}$.

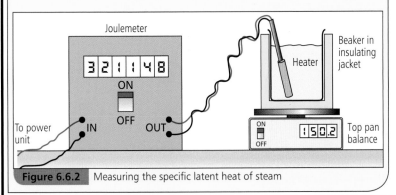

Figure 6.6.2 Measuring the specific latent heat of steam

KEY POINTS

1 Latent heat is energy used or released when a substance changes its state without changing its temperature.

2 Specific latent heat of fusion (or vaporisation) is the energy needed to melt (or boil) 1 kg of the substance without change of temperature.

SUMMARY QUESTIONS

1 In the experiment shown in Figure 6.6.1, 0.024 kg of water was collected in the beaker in 300 s with the heater off. The beaker was then emptied and placed under the funnel again. With the heater on for exactly 300 s, the joulemeter reading increased from zero to 15 000 J and 0.068 kg of water was collected in the beaker.

   a Calculate the mass of ice melted due to the heater being on.

   b Calculate the specific latent heat of ice.

2 a In the experiment shown in Figure 6.6.2, the balance reading decreased from 0.152 kg to 0.144 kg in the time taken to supply 18 400 J of energy to the boiling water. Calculate the specific latent heat of steam.

   b After the heater in **a** was switched off, the temperature of the water in the beaker gradually decreased. What does this tell you about the experiment and the value calculated for the specific latent heat of steam?

Heat transfer (1)
Thermal conduction

Supplement

LEARNING OUTCOMES

- Describe experiments to compare the thermal conduction of different materials

- Explain why metals are good conductors of heat and non-metals are poor conductors

When you are cooking food, you need to know which materials are good thermal conductors and which are good insulators. If you can't remember, you are likely to burn your fingers.

Comparing rods of different materials as conductors

The rods need to be the same width and length for a fair comparison. Each rod is coated with a thin layer of wax near one end. The uncoated ends are then heated together, as shown in Figure 6.7.1. The wax melts fastest on the rod that conducts best.

Rods of different metals and non-metals (e.g. glass) could be used in the test. The results show that:

- metals conduct better than non-metals
- copper is a better conductor than steel.

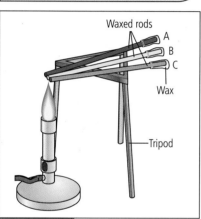

Figure 6.7.1 Comparing conductors

STUDY TIP

Notice the very important point made about the control of variables in the experiment.

PRACTICAL

Testing sheets of materials as insulators

1 Use different materials to insulate identical cans (or beakers) of hot water. Also, include a can without insulation. The volume of water and its temperature at the start should be the same.

2 Use a thermometer to measure the water temperature after the same time. The results should tell you which material is a good insulator and which insulator is best.

material	starting temperature	temperature after 300 s
no insulation	40 °C	22 °C
paper	40 °C	33 °C
felt	40 °C	38 °C

The results in the table show that:

- felt is a much better insulator than paper
- paper is better than no insulation.

DID YOU KNOW?

Good insulators

Materials like wool and fibre glass are good insulators. This is because they contain air trapped between the fibres. Trapped air is a good insulator. We use insulators such as fibre glass to cut down heat transfer:

- from hot objects to colder surroundings (e.g. loft insulation in buildings, lagging hot water pipes)
- to cold objects from hotter surroundings (e.g. food in a freezer or cool box).

Insulating a loft

Conductors in metals

Metals contain lots of **conduction** (or 'free') **electrons**. These electrons have broken free from the atoms, leaving each atom as a positively charged ion. Free electrons move about at random inside the metal and bond the positive ions together. They collide with each other and with the positive ions.

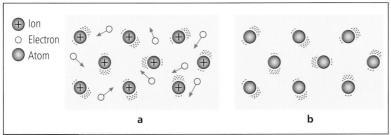

○+ Ion
○ Electron
● Atom

a   b

Figure 6.7.3   Energy transfer (a) in a metal, (b) in a non-metal

When a metal rod is heated at one end, the free electrons at the hot end gain kinetic energy and move faster.

• These electrons **diffuse** (i.e. spread out) and collide with other free electrons and the ions in the cooler parts of the metal.

• As a result, they transfer kinetic energy to these electrons and ions.

Energy is therefore transferred from the hot end of the rod to the colder end.

In a non-metallic solid, all the electrons are held in the atoms. There are no free electrons. Energy transfer takes place only because the atoms vibrate and shake each other. This is much less effective than energy transfer by free electrons. This is why metals are much better conductors than non-metals.

PRACTICAL

How well does water conduct heat?

Heat a test tube of water near the top with a 'weighted' ice cube near the bottom. Even when the water at the top starts boiling, the ice cube does not melt. Water is a poor conductor of heat.

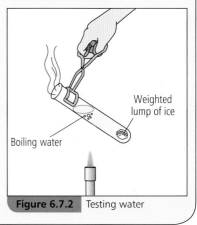

Weighted lump of ice

Boiling water

Figure 6.7.2   Testing water

STUDY TIP

You should be able to describe the process of thermal conduction in terms of conduction (or 'free') electrons.

SUMMARY QUESTIONS

1 Choose the best insulator or conductor from the list for a, b and c.

fibre glass   plastic   steel   wood

  a _____ is used to insulate a house loft.

  b The handle of a frying pan is made of _____ or _____.

  c A radiator in a central heating system is made from _____.

2 a Why does a woolly hat keep your head warm in cold weather?

  b Choose a material you would use to line the inside of a cool box to be used to keep food cool in summer? Explain your choice of material.

  c Explain why an ice cube in water melts faster if the water is stirred.

KEY POINTS

1 Materials such as fibre glass are good insulators because they contain pockets of trapped air.

2 Conduction in a metal is due to free electrons transferring energy inside the metal.

3 Non-metals are poor conductors because they do not contain free electrons.

Supplement

Heat transfer (2) Convection

LEARNING OUTCOMES

- Recall that convection happens in a gas or a liquid when it is heated
- Describe experiments that show convection
- Explain applications that use convection
- Relate convection to density changes

A natural glider

STUDY TIP

Make sure you are able to write a clear description of the process of convection.

A hot-air balloon carries its own burner

Observing convection

Glider pilots and birds know how to use convection to stay in the air. Convection currents of warm air can keep them high above the ground for hours.

Convection happens whenever a fluid (i.e. a gas or a liquid) is heated. The fluid expands where it is heated and becomes less dense. As a result, the heated fluid rises – just as objects in water float if they are less dense than water. As the fluid rises, it mixes with colder fluid and becomes cooler. The convection currents in a fluid transfer energy from the hotter parts to the cooler parts.

Figure 6.8.1 shows a simple demonstration of convection. The hot gases from the burning candle go straight up the chimney above the candle. The hot gases are replaced by cold air drawn down the other chimney to replace the air leaving the box.

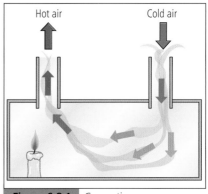

Figure 6.8.1 Convection

Using convection

• Hot water at home

Many homes have a hot water tank. Hot water from the boiler rises and flows into the tank where it rises to the top. Figure 6.8.2 shows the system. When you use a hot water tap at home, you draw hot water from the top of the tank.

• Hot air for fun

A hot air balloon carries its own burner to heat the air in the balloon. This makes air in the balloon less dense than the surrounding air. So the balloon rises and floats in the air. To keep the balloon floating, the air in the balloon needs to be heated every so often. If this isn't done, the air cools and the balloon descends.

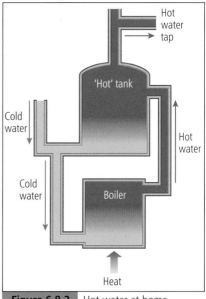

Figure 6.8.2 Hot water at home

Sea breezes

Sea breezes keep you cool on a beach in hot weather. On a sunny day, the ground heats up faster than the sea. So the air above the ground warms up and rises. Cooler air from the sea flows in as a sea breeze to take the place of the warmer air.

How convection works

Convection takes place:

• only in liquids and gases (i.e. fluids)
• due to circulation (convection) currents within the substance.

The circulation currents are caused because fluids rise where they are heated (as heating makes them less dense) and fall where they cool (as cooling makes them more dense). Convection currents transfer thermal energy from the hotter parts to the cooler parts.

Why do liquids and gases rise when heated? Most liquids and gases expand when heated. This is because the particles move about more and take up more volume. Therefore the density decreases because the same mass of liquid or gas occupies a bigger volume. So heating a liquid or gas makes it less dense and it therefore rises.

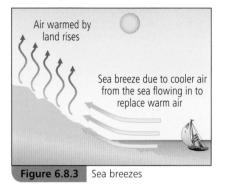

Air warmed by land rises

Sea breeze due to cooler air from the sea flowing in to replace warm air

Figure 6.8.3 Sea breezes

STUDY TIP

Don't make vague, inaccurate statements like 'heat rises'. It is the hot gas or hot liquid that rises.

SUMMARY QUESTIONS

1 Use each word from the list once only to copy and complete the following sentences.

cools   falls   mixes   rises

When a fluid is heated, it _____ and _____ with the rest of the fluid. The fluid circulates and _____ then it _____.

2 Figure 6.8.4 shows a convector heater. It has an electric heating element inside and a metal grille on top.

   a What does the heater do to the air inside the heater?

   b Why is there a metal grille on top of the heater?

   c Where does air flow into the heater?

Hot air

Figure 6.8.4

3 Some coloured crystals are placed on the bottom of a water-filled beaker. When the water is heated as shown in Figure 6.8.5, streams of coloured water are seen to rise from the crystals and travel across the surface and down the side of the beaker. Explain this observation.

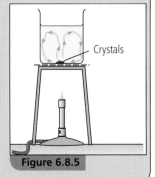

Crystals

Figure 6.8.5

KEY POINTS

1 Convection takes place only in liquids and gases.

2 Heating a liquid or a gas makes it less dense.

3 Convection is due to a hot liquid or gas rising.

Supplement

LEARNING OUTCOMES

- Identify infra-red radiation as part of the spectrum of electromagnetic radiation
- Recognise that all objects emit infra-red radiation which increases with increasing temperature
- Describe experiments which show the properties of good and bad emitters and absorbers of infra-red radiation

Seeing in the dark

We can use special TV cameras to 'see' animals and people in the dark. These cameras detect infra-red radiation. Every object around us emits infra-red radiation. The hotter an object is, the more infra-red radiation it emits. If an object is very hot, it emits light as well. Light and infra-red radiation are part of the spectrum of electromagnetic radiation. So too are radio waves, microwaves, ultraviolet rays and X-rays.

To detect infra-red radiation, we can use a thermometer with a blackened bulb. Figure 6.9.1 shows how to do this.

- The glass prism splits a narrow beam of light from a lamp bulb into the colours of the spectrum.
- The thermometer reading rises when it is placed just beyond the red part of the spectrum. Some of the infra-red radiation from the lamp bulb goes there. It is called infra-red radiation because it is next to the red part of the visible spectrum.
- Our eyes cannot detect infra-red radiation; the thermometer can.

The radiation emitted by an object due to its temperature is sometimes called **thermal radiation**. This is because it can include light as well as infra-red radiation if the object is hot enough.

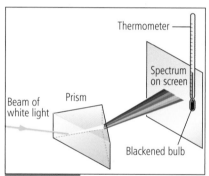

Figure 6.9.1   Detecting infra-red radiation

Surfaces and radiation

Which surfaces are the best emitters of infra-red radiation?

Rescue teams use thermal blankets to keep accident survivors warm. A thermal blanket has a shiny white outer surface. This emits much less radiation than a dull black surface. The colour and smoothness of a surface affects how much radiation it emits.

Dull black surfaces emit infra-red radiation better than shiny white surfaces.

Which surfaces are the best absorbers of infra-red radiation?

When a photocopier is used, the copies are warm. This is because infra-red radiation from a lamp is used to dry the ink on the paper. Otherwise, the copies will be smudged. The black ink absorbs infra-red radiation more easily than the white paper.

- A black surface absorbs infra-red radiation better than a white surface.
- A dull surface absorbs infra-red radiation better than a shiny surface because it has lots of cavities. Figure 6.9.2 shows why these cavities trap and absorb the infra-red radiation.

Dull black surfaces absorb infra-red radiation better than shiny white surfaces.

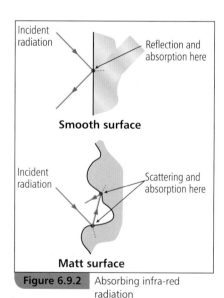

Figure 6.9.2   Absorbing infra-red radiation

PRACTICAL

Testing different surfaces

1 **To compare the radiation from two different surfaces**, measure how fast two beakers (or cans) of hot water cool. One beaker needs to be shiny and silvery and the other dull black. Figure 6.9.3 shows the idea. At the start, the volume and temperature of the water in each beaker need to be the same. The dull black beaker cools faster than the other one.

2 **To compare absorption by different surfaces**, use two beakers as shown with cold water in. Place the beakers in a sunlit room in the sunlight.

You should find that a beaker with a shiny, silvery surface warms up slower than a beaker painted dull black.

3 In these tests, the two cans must have equal surface areas and be at the same initial temperatures, as well as containing equal volumes of water. This is because the amount of radiation emitted from an object depends on the **area** and **temperature** of its surface, as well as on the surface colour and how shiny or dull it is.

Figure 6.9.3 Testing different surfaces

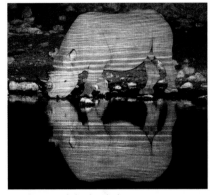

Keeping watch in the darkness

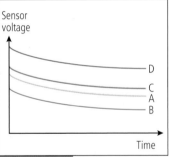

Figure 6.9.5 Graph for Summary Question 2

SUMMARY QUESTIONS

1 Explain why:
  a penguins huddle together to keep warm,
  b houses in hot countries are usually painted white,
  c solar heating panels are painted black.

2 A metal cube filled with hot water was used to compare the heat radiated from its four vertical faces, A, B, C and D.

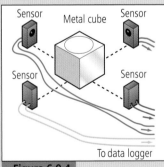

Figure 6.9.4

An infra-red sensor was placed opposite each face at the same distance, as shown in Figure 6.9.4. The more intense the infra-red radiation detected by a sensor, the greater its output voltage was. The sensors were connected to a computer. The results of the test are shown in the graph in Figure 6.9.5.

  a Which face radiated **i** most? **ii** least?
  b Why was it important for the distance from each sensor to the face to be the same?
  c One face was white and shiny, one was white and dull, one was black and shiny and one was dull black.

  Which face was **i** white and shiny? **ii** dull and black?

KEY POINTS

1 Infra-red radiation is part of the spectrum of electromagnetic radiation.

2 All bodies emit infra-red radiation.

3 Dull black surfaces are better emitters and absorbers of infra-red radiation than shiny white surfaces.

4 The hotter a body is, the more infra-red radiation it emits. The greater the surface temperature or the surface area, the more infra-red radiation it emits.

LEARNING OUTCOMES

- Recognise that we need to increase heat transfer to stop devices like engines from over-heating
- Recall the main methods of increasing heat transfer
- Recognise that we need to reduce heat transfer to keep objects warm
- Explain the main methods of reducing heat transfer

A metal heat sink

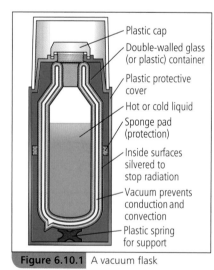

- Plastic cap
- Double-walled glass (or plastic) container
- Plastic protective cover
- Hot or cold liquid
- Sponge pad (protection)
- Inside surfaces silvered to stop radiation
- Vacuum prevents conduction and convection
- Plastic spring for support

Figure 6.10.1 A vacuum flask

Cooling by design

Lots of things can go wrong if heat transfer isn't controlled. A car engine that overheats can go up in flames. An electronic component in an electrical device that overheats will stop working and may cause the device to malfunction.

A vehicle radiator is designed to transfer heat energy from the engine to the surroundings.

Most vehicles have a water-cooled engine. Water pumped through channels in the engine case is heated by the engine before flowing through the radiator. As the hot water

A car radiator

flows through the radiator, it loses heat energy as a result of heat conduction through the metal case of the radiator. The radiator is flat, so it has a large surface area. This increases the amount of air heated by the radiator, so increasing the heat loss through convection in the air and through radiation. Most cars also have a cooling fan that switches on when the engine is too hot. This increases the flow of air over the radiator surface.

A heat sink is designed to stop an electronic component from overheating. By attaching a component that might overheat to a metal heat sink, the effective surface area of the component is increased. This increases heat losses due to convection and radiation. In addition, if the heat sink has a dull black surface, the heat losses due to radiation are further increased.

Keeping warm

The vacuum flask

If you are outdoors in cold weather, a hot drink from a vacuum flask keeps you warm. The same flask can also be used in hot weather to keep a cold drink cold.

The liquid is in the double-walled glass or plastic container.

- The vacuum between the two walls of the container cuts out heat transfer by conduction and convection between the walls. However infra-red radiation travels through a vacuum as it is electromagnetic radiation.
- Glass is a poor conductor so there is little heat conduction through the glass.
- The glass surfaces are silvery to reduce radiation from the outer wall.

Reducing heat transfer at home

In cold weather, home heating bills can be expensive. Figure 6.10.2 shows how people in cold climates can reduce heat losses in cold weather in their homes and cut home heating bills.

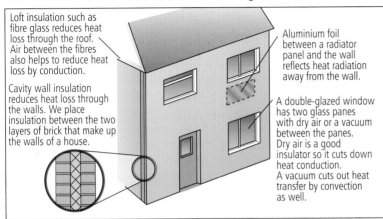

Loft insulation such as fibre glass reduces heat loss through the roof. Air between the fibres also helps to reduce heat loss by conduction.

Cavity wall insulation reduces heat loss through the walls. We place insulation between the two layers of brick that make up the walls of a house.

Aluminium foil between a radiator panel and the wall reflects heat radiation away from the wall.

A double-glazed window has two glass panes with dry air or a vacuum between the panes. Dry air is a good insulator so it cuts down heat conduction. A vacuum cuts out heat transfer by convection as well.

Figure 6.10.2

On a hot day, heat transfer into buildings from outside needs to be reduced. Air-conditioning units keep the interior of buildings cool but they are expensive to run. Measures to reduce heat transfer into a building would reduce the need for air-conditioning units. Such measures include:

- painting outside walls and roofs white so they absorb less heat from the Sun because they would reflect sunlight better than dark buildings
- fitting white blinds or window shutters to reflect sunlight and thus stop rooms becoming too hot
- building thick walls (when the building is constructed) to cut down heat transfer due to conduction through the walls.

STUDY TIP

Questions in the examination may ask about cooling or keeping warm. You must decide whether the object you are asked about is at a temperature above or below the surroundings and then use the correct description about heat emission or absorption.

SUMMARY QUESTIONS

1 Hot water is pumped through a central heating radiator like the one in Figure 6.10.3.

Copy and complete the sentences below about the radiator.

a Heat transfer through the walls of the radiator is due to _____.

b Hot air in contact with the radiator causes heat transfer to the room by _____.

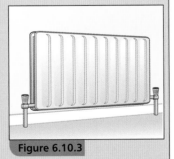

Figure 6.10.3

2 a A double-glazed window has a plastic frame and a vacuum between the panes.

  i Why is a plastic frame better than a metal frame?

  ii Why is a vacuum between the panes better than air?

 b Give two reasons why a building with thick external walls heats up less in hot weather than an identical building with thinner walls.

KEY POINTS

1 Heat transfer to or from an object can be increased by increasing the surface area of the object or making its surface dull black.

2 Heat transfer in cold weather from a building can be reduced using aluminium foil behind radiators, cavity wall insulation, double glazing and loft insulation.

3 Heat transfer in summer into a building can be reduced by painting outdoor surfaces white, using white window shutters or blinds, and having thick walls.

1 The photograph shows an expansion gap in a motorway.

(a) Describe what could happen if expansion gaps were not included in the design of the road.

(b) Explain why the expansion gap works.

2 (a) State the temperatures of the following:

   (i) boiling point of water

   (ii) freezing point of water.

(b) Suggest a typical temperature for each of the following:

   (i) room temperature

   (ii) air temperature on a very hot day

   (iii) normal body temperature.

3 The graph shows the results obtained by a student who was heating a solid until it melted.

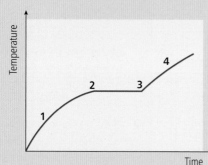

(a) Describe what is happening to the substance that is being heated:

   (i) at point 1 on the graph

   (ii) between 2 and 3 on the graph

   (iii) at point 4 on the graph.

(b) Copy the sketch graph and mark on the temperature axis the position of the melting point of the solid.

(c) Explain in terms of the movement of molecules what is happening:

1 The reading in °C shown on the thermometer is approximately

A 10.2   **B** 12   **C** 20.8   **D** 28

(Paper 1/2)

2 The average body temperature in °C for a healthy person is approximately

A 20   **B** 28   **C** 37   **D** 42

(Paper 1/2)

3 Which of the following best describes the heat radiation properties of a matt black surface?

  A good absorber and good emitter

  B good absorber and poor emitter

  C poor absorber and good emitter

  D poor absorber and poor emitter

(Paper 1/2)

4 The unit of specific heat capacity is

  A J   **B** J/°C   **C** J/kg   **D** J/kg°C

(Paper 2)

   (i) between 1 and 2 on the graph

   (ii) between 2 and 3 on the graph

   (iii) between 3 and 4 on the graph.

4 An object is supplied with 6000 J of heat energy and its temperature increases from 15 °C to 18 °C. Calculate the thermal capacity of the object.

5 (a) Define the term specific heat capacity.

(b) A block of metal of mass 800 g is heated from 20 °C to a temperature of 35 °C. The amount of heat energy supplied is 9600 J. Calculate the specific heat capacity of the metal.

6 When choosing the type of thermometer to use for particular circumstances a number of factors should be considered. Explain briefly what is meant by the following terms.

(a) Range of temperatures to be measured

(b) Sensitivity of a thermometer

(c) Linearity of a thermometer

(d) Accuracy of a thermometer

Supplement

5 The property of a thermometer that is described as the change of the thermometric property for a 1°C rise in temperature is called

A accuracy   **C** range

B linearity   **D** sensitivity

(Paper 2)

6 The diagram shows the apparatus used for an experiment to find out which of three materials is the best conductor of heat.

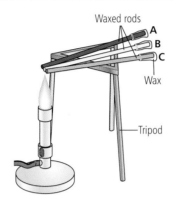

Waxed rods
A
B
C
Wax
Tripod

(a) (i) State what you would expect to happen to the wax when the rods are heated.

(ii) Explain how the result would tell you which of the three rods is made from the best conducting material.

(iii) Suggest two possible variables that must be kept constant in order to obtain a meaningful result.   *[5]*

(b) Copy and complete the sentences:

A saucepan is made from steel but has a wooden handle.

Steel is used because it is a _____.

Wood is used because it is a _____ and so prevents your hand being _____.   *[3]*

(Paper 3)

7 The diagram shows a room heater.

(a) Copy and complete the sentences to describe how the room is heated.

Air above the heater is warmed and this makes the air _____ . As a result the _____ of the air is reduced and the warm air _____ . Cold air takes its place which is then warmed. In this way _____ currents are set up and the whole room becomes warm.   *[4]*

(b) (i) Infra-red radiation is part of the electromagnetic spectrum. Name two other types of wave that are part of the electromagnetic spectrum.

(ii) Thermal blankets are used to keep accident survivors warm. The blankets have a light coloured shiny surface. Explain why such a surface is important.   *[4]*

(Paper 3)

8 The graph shows the results for a cooling experiment. A substance was heated until it melted and then its temperature was recorded as it was allowed to cool back to room temperature.

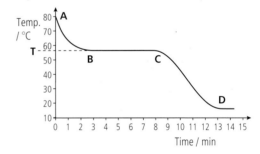

Temp. / °C

(a) (i) State the significance of the temperature marked T on the graph.

(ii) Describe the state of the substance in regions A–B and C–D on the graph.

(iii) Explain in terms of the movement of molecules what is happening to the substance in region B–C of the graph.

(iv) Use the graph to estimate a value for room temperature.   *[6]*

(b) A second sample of the substance was heated until it boiled. While it was liquid, but before it began to boil, some of the liquid evaporated. Explain the differences between evaporation and boiling.   *[4]*

(Paper 4)

7.1   Wave motion

Supplement

LEARNING OUTCOMES

- Describe and observe the wave motion of different types of waves
- Explain the meaning of speed, frequency, wavelength and amplitude
- Use the equation 'speed = frequency × wavelength'

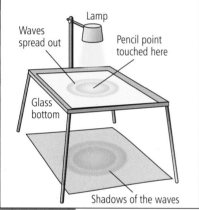

Figure 7.1.1   The ripple tank

Lamp
Waves spread out
Pencil point touched here
Glass bottom
Shadows of the waves

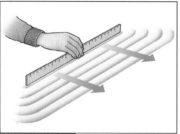

Figure 7.1.2   Making water waves

STUDY TIP

A common error is to think that the amplitude of a wave is the distance from the top of a crest to the bottom of the next trough.

To understand wave motion, we can study waves on water and waves on ropes and on springs. Then we can apply what we learn to any wave motion. All the examples below show waves as repeated disturbances that travel along a rope or spring or across water. The examples show that waves transfer energy without transferring matter.

Water waves can be produced in many ways. If you drop a pebble in a pond, the waves spread out in expanding circles. The expanding circles are called **wavefronts.** Anything floating on the water will bob up and down as successive wave fronts pass by. Figure 7.1.1 shows how we can study water waves in controlled conditions using a ripple tank. Touch the centre of the water surface and you will see the waves spread out in all directions.

PRACTICAL

Making straight waves

1 Make straight waves by moving a rule up and down on the water surface in a ripple tank (Figure 7.1.2). The wavefronts travel away from the rule. The straight wavefronts are called plane waves. They move at the same speed and they keep the same distance apart.

2 Observe the effect on the waves of moving the rule up and down faster then slower. More wavefronts are produced every minute and they become closer together. Use a stopwatch to find out if their speed has changed.

Waves on a rope can be produced by moving one end of the rope from side to side repeatedly. This action sends waves along it. Each part of the rope vibrates from side to side. The strings of musical instruments vibrate in this way to create sound waves in air.

Waves on a long stretched spring, such as a slinky coil, can be produced by fixing one end of the spring and moving the other end repeatedly to and fro in any direction. The motion of this part of the spring creates waves that travel along the spring. Each part of the spring vibrates and makes the next part vibrate, which makes the next part vibrate, and so on. So the vibrations travel along the spring.

Measuring waves

The speed of a wave, v, is the distance travelled by a wave crest or a wave trough every second. For example, sound waves in air travel at a speed of 340 m/s. In 5 s, sound waves travel a distance of 1700 m (= 340 m/s × 5 s).

Figure 7.1.3 shows a snapshot of waves travelling along a rope.

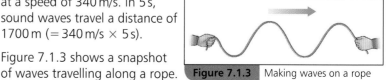

Figure 7.1.3   Making waves on a rope

Each point on the rope vibrates up and down. The words we use to describe waves are shown in Figure 7.1.4 below.

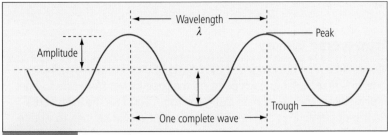

Figure 7.1.4 Features of a wave

- **One complete wave** is from one wave peak to the next.
- **The frequency, *f*,** of the waves is the number of complete waves (or wave peaks) passing a point in one second. The unit of frequency is the hertz (Hz) named after Heinrich Hertz, the scientist who discovered how to produce and detect radio waves.
- **The wavelength, λ** (Greek, pronounced 'lambda'), of the waves is the distance from one wave peak to the next.
- **The amplitude** of the waves is the height of the wave peak or the depth of the wave trough from the middle. The bigger the amplitude of the waves, the more energy the waves carry.

Supplement

The speed of the waves depends on the frequency and the wavelength according to the following equation:

speed = frequency × wavelength

$$v = f\lambda$$

The proof of this equation is not required in your examination. However, awareness of the proof gives a deeper understanding of the link between frequency, speed and wavelength.

Figure 7.1.5 shows a surfer riding on the crest of some unusually fast waves. Suppose the frequency of the waves is 3 Hz and the wavelength of the waves is 4 m. At this frequency, three wave crests pass a fixed point once every second. The surfer on a wave crest therefore moves forward a

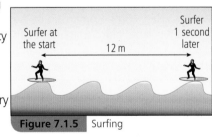

Figure 7.1.5 Surfing

distance of 3 wavelengths every second, and therefore in 1 s travels a distance of 12 m (= 3 wavelengths × 4 m for each wavelength). So the speed of the surfer is 12 m/s, which is equal to the frequency × the wavelength of the waves.

1 Copy and complete the following sentences using words from the list below.

amplitude frequency speed wavelength

a The hertz is the unit of _____.

b The distance from one wave peak to the next is the _____ of a wave.

c For water waves, the height of a wave crest above the undisturbed water surface is the _____ of the wave.

d _____ × frequency = _____.

2 Figure 7.1.6 shows a snapshot of a wave travelling from left to right along a rope.

Figure 7.1.6

a Mark on the diagram **i** one wavelength, **ii** the amplitude of the waves.

b Describe the motion of point P on the rope.

c **i** When the frequency of the waves is 2.0 Hz, their wavelength is 2.5 m. Calculate the speed of the waves.

   ii Assuming the speed is unchanged, calculate the wavelength of waves of frequency 10 Hz.

Supplement

KEY POINTS

1 Waves are repeated disturbances that move along.

2 The amplitude of a wave is the height of the wave crest or the depth of the wave trough from the middle.

3 Frequency is the number of wave crests passing a point in one second.

4 Wave speed = frequency × wavelength

Supplement

Supplement

LEARNING OUTCOMES

- Describe the differences between transverse waves and longitudinal waves
- State examples and describe applications of transverse and longitudinal waves

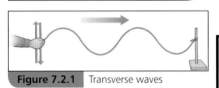

Figure 7.2.1 Transverse waves

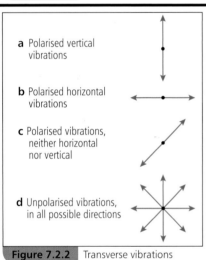

a Polarised vertical vibrations

b Polarised horizontal vibrations

c Polarised vibrations, neither horizontal nor vertical

d Unpolarised vibrations, in all possible directions

Figure 7.2.2 Transverse vibrations

PRACTICAL

1 Work in a small group. One person makes waves on a rope, another person holds a letter box slot in a board which the rope passes through.

2 Make unpolarised transverse waves travel along a rope by moving one end of the rope side-to-side with the other end fixed on the other side of the slot. Figure 7.2.3 shows the idea. You should find only those waves that vibrate in the same direction as the slot pass through it. The slot polarises the waves. The same effect happens when unpolarised light is directed at a Polaroid filter. The light waves that pass through the filter are polarised.

Transverse waves

Waves on a rope are produced by moving one end of the rope from side to side repeatedly. Each point on the rope moves from side to side repeatedly as the waves pass along the rope. Suppose the rope had a dark spot painted on it at one point. The spot would be seen to vibrate perpendicularly (i.e. at right angles) to the direction which the waves are moving. Such waves are called **transverse waves**. All electromagnetic waves are transverse waves. Thus light waves are transverse waves.

The vibrations of a transverse wave are perpendicular to the direction in which the waves are travelling.

Tranverse waves are said to be **polarised** if the vibrations are always along the same line (at right angles to the direction of travel of the waves). For example:

- in Figure 7.2.1, each point on the rope vibrates in a vertical line only. Figure 7.2.2a shows how the rope would look if it was viewed end-on.
- Figure 7.2.2b shows an end-view of a rope made to vibrate horizontally by moving one end from side to side along a horizontal line.
- Figure 7.2.2c shows an end-view of the rope if one end is moved from side to side along a line which is neither horizontal nor vertical. At any point on the rope, the vibrations are always along this line.

All three examples are polarised transverse waves because in each case, the vibrations are always along the same line. Figure 7.2.2d shows an end-view of **unpolarised** waves on a rope. These are created by continually changing the direction of vibration.

Light waves from a lamp bulb are unpolarised transverse waves. They can be polarised by passing them through special material called Polaroid. Its molecules are all lined up in the same direction. As a result, the material only allows light waves through that vibrate in the same direction as the molecules.

Figure 7.2.3 also shows what happens if the rope passes through a second slot at right angles to the first one. The polarised waves from the first slot cannot pass through the second slot. The same effect happens if polarised light is passed through a Polaroid filter which has its molecules lined up perpendicular to the vibration of the light waves.

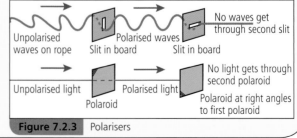

Figure 7.2.3 Polarisers

Supplement

Applications

1 **Polaroid filters** are used in liquid crystal displays (LCDs) which are widely used in TV screens, in computer monitors and in calculators. They are also used in Polaroid sunglasses to eliminate glare.

2 **Aerials** of radio receivers need to be lined up suitably to detect the signal. This is because the signal is carried by polarised radio waves. The aerial needs to be lined up with the vibrations of the radio waves.

Longitudinal waves

Sound waves and compression waves on a spring are longitudinal waves. The direction of the vibrations is parallel to the direction in which the waves travel. Because the vibrations are not transverse, longitudinal waves cannot be polarised.

The vibrations of a longitudinal wave are parallel to the direction in which the waves are travelling.

A **'slinky spring'** is useful for demonstrating how sound waves travel. If one end of the slinky is moved to and fro as shown in Figure 7.2.4, compression waves travel along the slinky. Each part of the slinky vibrates along the line of the spring. In other words, the vibrations are parallel to the direction in which the waves travel.

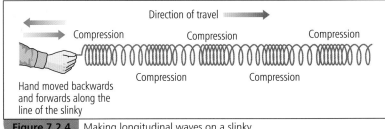

Figure 7.2.4 Making longitudinal waves on a slinky

Sound waves are created in air when an object vibrates. Its vibrating surface sends pressure waves through the surrounding air as the surface pushes and pulls repeatedly on the air. Figure 7.2.5 shows how sound waves are created by a loudspeaker. Each layer of air repeatedly moves back and forth along the direction of travel of the sound waves. In other word, the vibrations are parallel to the direction of travel of the sound wave. As each layer of air vibrates, it makes the next layer of air vibrate, so the vibrations move through the air.

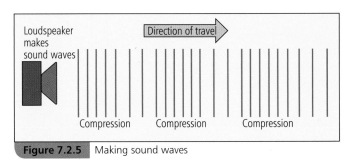

Figure 7.2.5 Making sound waves

KEY POINTS

1 Transverse waves vibrate at right angles to the direction of travel of the waves.

2 Longitudinal waves vibrate parallel to the direction of travel of the waves.

3 Electromagnetic waves and waves on a rope are transverse waves.

4 Sound waves and compression waves along a slinky spring are longitudinal waves.

SUMMARY QUESTIONS

1 Copy and complete the following sentences using words from the list below.

**longitudinal   parallel
transverse   perpendicular**

a Sound waves are _____ waves.

b Light waves are _____ waves.

c The vibrations of longitudinal waves are _____ to the direction in which the waves are travelling.

2 a Describe how you would use a slinky spring to demonstrate to a friend the difference between longitudinal and transverse waves.

b If a beam of unpolarised light is passed through two Polaroid filters, the beam can be stopped by turning the second filter. Explain why this happens.

Supplement

Wave properties – reflection and refraction

LEARNING OUTCOMES

- Describe the reflection and refraction of plane waves in a ripple tank
- Recall that refraction is due to a change in the speed of waves when they cross a boundary

Investigating waves using a ripple tank

Reflection of plane waves can be investigated using the ripple tank shown in Topic 7.1. Plane (i.e. straight) waves produced by dipping a rule in water are directed at a metal barrier in the water. These waves are referred to as the incident waves to distinguish them from the reflected waves. The incident waves are reflected by the barrier. Figures 7.3.1a and b each show a wavefront before and after hitting the barrier.

- In Figure 7.3.1a, the incident wavefront is parallel to the barrier as it approaches the barrier. It is still parallel to the barrier after reflection as it travels away from the barrier.
- In Figure 7.3.1b, the incident wavefront is not parallel to the barrier before or after reflection. The reflected wavefront moves away from the barrier at the same angle to the barrier as the incident wavefront.

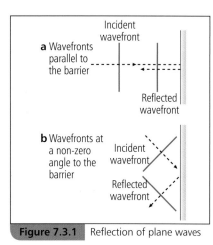

Figure 7.3.1 Reflection of plane waves

PRACTICAL

A reflection test using a ripple tank

Use a rule to create and direct plane waves at a straight barrier, as shown in Figure 7.3.1. Find out if the reflected waves are always at the same angle to the barrier as the incident waves. You could align a second rule with the reflected waves and measure the angle of each rule to the barrier. Repeat the test for different angles.

Refraction of waves is the change of the direction in which they are travelling when they cross a boundary. This can be observed in a ripple tank when water waves cross a boundary between 'deep' and 'shallow' water. An area of shallow water can be created by placing a glass plate flat in the water. The water above the plate is shallower than the water outside the plate area. Plane waves are directed at a non-zero angle to a boundary. The wavefronts change direction as they cross the boundary, as shown in Figure 7.3.2.

PRACTICAL

Refraction tests

1   Use a vibrating beam to create plane waves continuously in a ripple tank containing a transparent plastic plate. Arrange the plate so the waves cross a boundary between the deep and shallow water.

2   **At a non-zero angle to a boundary**, as shown in Figure 7.3.2, the waves bend when they cross the boundary. The water over the plate needs to be very, very shallow. Find out if plane waves bend towards or away from the boundary when they cross from deep to shallow water.

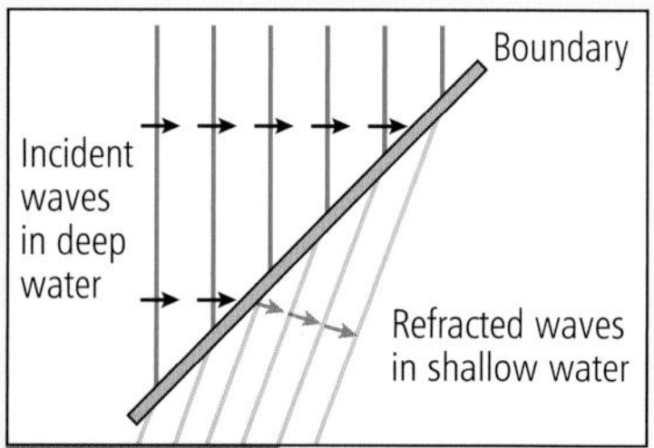

Figure 7.3.2 Observing refraction

3   **Parallel to a boundary**. The waves cross the boundary without bending or changing direction. However, their speed changes. Find out if the waves travel slower or faster when they cross the boundary. As we shall see later, this change of speed explains why the waves are refracted when they cross a boundary at a non-zero angle.

Describing reflection and refraction

To explain how a wavefront moves forward, imagine each tiny section creates a wavelet which travels forwards. The wavelets move forward together to recreate the wavefront that created them.

Refraction: when plane waves cross a boundary at a non-zero angle to the boundary where they slow down, each wavefront changes its direction.

In Figure 7.3.3, the wavefronts move more slowly after they have crossed the boundary. They are not as far from the boundary as they would have been if their speed had not changed. So the refracted wavefronts are at a smaller angle to the boundary than the incident wavefronts.

Reflection: when plane waves reflect from a flat barrier, the reflected waves are at the same angle to the barrier as the incident waves. When each point on the wavefront reaches the barrier, it creates a wavelet moving away from the barrier. This wavelet lines up with the previous 'reflected' wavelets to form a reflected wavefront moving away from the barrier. All parts of a wavefront move at the same speed. This means that the reflected wavefront is at the same angle to the barrier as the incident wavefront.

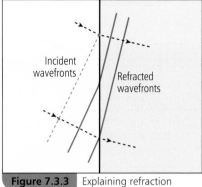

Figure 7.3.3 Explaining refraction

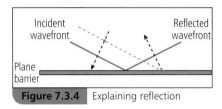

Figure 7.3.4 Explaining reflection

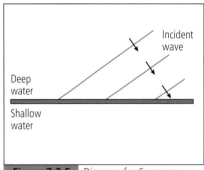

Figure 7.3.5 Diagram for Summary Question 2

SUMMARY QUESTIONS

1 Copy and complete the following sentences using words from the list below.

more than   less than   the same as

a When plane waves reflect from a straight barrier, the angle of each reflected wave to the barrier is _____ the angle of each incident wave to the barrier.

b When waves are refracted, the angle of each refracted wavefront is not _____ the angle of each incident wave front to the barrier.

c When waves speed up on crossing a boundary, the angle of each refracted wavefront is _____ the angle of each incident wave front to the barrier.

2 Copy Figure 7.3.5 which shows plane waves passing from deep to shallow water, where they move more slowly than in the deep water. Draw in some refracted wavefronts, indicating their direction.

3 Sea waves rolling up a sandy beach are not reflected. Why are the sides of a ripple tank sloped? What would happen if the sides were vertical rather than sloped?

KEY POINTS

1 Plane waves reflect from a straight barrier at the same angle to the barrier as the incident waves.

2 Refraction is the change of direction of waves when they cross a boundary due to a change in the speed of waves when they cross a boundary.

Wave properties – diffraction

Supplement

LEARNING OUTCOMES

- Describe diffraction of plane waves in a ripple tank
- Recall the effect on gap width and wavelength on the diffraction of waves passing through a gap
- Recall the effect of wavelength on the diffraction of waves passing the edge of an obstacle

STUDY TIP

The definition of diffraction is simple, but you must learn it.

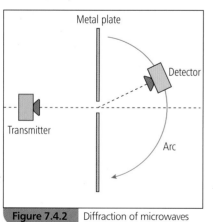

A Hubble Space Telescope image of two colliding galaxies

Figure 7.4.2 Diffraction of microwaves (top view)

Metal plate
Detector
Transmitter
Arc

Investigating diffraction using a ripple tank

Diffraction is the spreading of waves when they pass through a gap or move past an obstacle. The waves that pass the edges of the gap or of the obstacle spread out. Figure 7.4.1 shows waves in a ripple tank spreading out after they pass through a gap. Notice that:

- the narrower the gap, the more the waves spread out
- the wider the gap, the less the waves spread out.

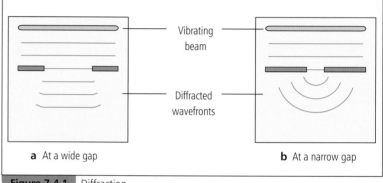

Vibrating beam

Diffracted wavefronts

a At a wide gap   **b** At a narrow gap

Figure 7.4.1 Diffraction

Diffraction is important in any optical instrument. The Hubble Space Telescope in its orbit above the Earth has provided amazing images of objects far away in space. Its focusing mirror is 2.4 m in diameter. When it is used, astronomers can see separate images of objects which are far too close to be seen separately using a narrower telescope. Little diffraction occurs when light passes through the Hubble Space Telescope because it is so wide. So its images are very clear and very detailed.

PRACTICAL

Investigating diffraction

Use a ripple tank as in Figure 7.4.1 to direct plane waves continuously at a gap between two metal barriers.

1 Change the gap spacing and observe the effect on the diffraction of the waves that pass through the gap. You should find that the diffraction of the waves increases as the gap is made narrower, as shown in Figure 7.4.1.

2 Keep the gap spacing constant and change the wavelength of the waves by altering the frequency of the vibrating beam. Observe the effect on the diffraction of the waves. You should find that the smaller the wavelength of the waves, the more they are diffracted.

Tests using microwaves

A microwave transmitter and a detector can be used to demonstrate diffraction of microwaves. The transmitter produces microwaves of wavelength 3.0 cm. The detector includes a meter which gives a non-zero reading when it detects microwaves.

1   The detector is placed in the path of the microwave beam from the transmitter. When a metal plate is placed between the transmitter and the detector, the meter reading drops to zero. This shows that microwaves cannot pass through metal.

2   Two metal plates with a narrow gap between them are placed in the path of the beam from the transmitter. When the detector is moved along an arc at a constant distance beyond the gap, as in Figure 7.4.2, it detects microwaves that have spread out from the gap. When the gap is made wider, the microwaves passing through the gap spread out less. The detector needs to be nearer the centre of the arc to detect the microwaves.

Supplement

Diffraction factors

When waves pass through a gap, they spread out. For diffraction at a gap to be noticeable, the gap must be similar in width to the wavelength of the waves. The waves spread out more if:

• the gap is made narrow, as shown in Figure 7.4.1
• the wavelength of the waves is increased, as shown in Figure 7.4.3.

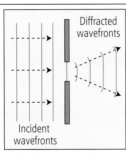
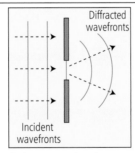

a Smaller wavelength than **b** with the same gap width

b Longer wavelength than **a** with the same gap width

Figure 7.4.3   The effect of wavelength on diffraction and wavefronts

When waves pass near an edge of an obstacle, they diffract and spread out behind the obstacle, as shown in Figure 7.4.4. The waves spread out more if the wavelength of the waves is increased.

Notice that the wavelength does not change when diffraction occurs.

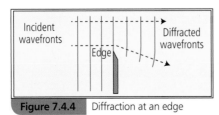

Figure 7.4.4   Diffraction at an edge

SUMMARY QUESTIONS

1 a   A small portable radio in a room can be heard all along a corridor outside the room when the room door is open. Explain why it can be heard by someone in the corridor who is not near the door.

  b   State what is meant by diffraction.

2   Copy and complete the following sentences about waves passing through a gap using words from the list below.

  more than   less than the same as

  a   Diffracted waves spread out _____ they would if the gap was made wider.

  b   When waves pass through a gap, their wavelength is _____ it was before it passed through the gap.

KEY POINTS

1   Diffraction is the spreading of waves when they pass through a gap or around an obstacle. For noticeable diffraction the gap must be similar in size to the wavelength.

2   The wavelength does not change on diffraction.

3   The narrower the gap, the more the waves spread out.

4   When waves pass through a gap, or an edge, the larger the wavelength, the more they diffract.

Supplement

1 When a wave is **(a)** refracted or **(b)** diffracted, which of the following, if any, will change?
   - the speed of the waves
   - the frequency of the waves
   - the wavelength of the waves.

2 During a thunderstorm a student sees a flash of lightning and hears the thunder 4 seconds later. If the speed of sound in air is 330 m/s, calculate how far the storm is from the student.

3 Draw diagrams to illustrate a transverse wave and a longitudinal wave.

   On each diagram mark a distance of one wavelength.

4 Write down three differences between sound waves and radio waves.

5 A microwave oven has a glass door with a metal grid. The grid has a large number of small holes. Explain why, when the door is closed and the oven switched on, you can see into the oven but microwaves cannot get out.

6 A ball is floating in a pond. The ball is out of reach. A person tries to make the ball move to one side of the pond by hitting the water with a stick and making waves.

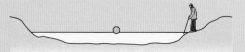

   Explain whether or not this is an effective way to move the ball across the pond.

Supplement

7 The person in question **6** hits the water at regular intervals 10 times in 20 s. The waves produced travel across the water at 0.5 m/s. Calculate the wavelength of the waves. Show your working.

1 The unit of frequency is
   A   Hz
   B   m
   C   m/s
   D   s
   (Paper 1/2)

2 The diagram shows a transverse wave.

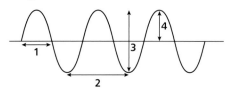

   On the diagram the amplitude is labelled
   A   1
   B   2
   C   3
   D   4
   (Paper 1/2)

3 The word 'diffraction' refers to
   A   dispersing of waves
   B   reflecting of waves
   C   refracting of waves
   D   spreading of waves
   (Paper 1/2)

4 Which of the following are examples of transverse waves?

   1 sound waves   2 water waves

   3 microwaves   4 radio waves

   A   1 only
   B   1 and 2 only
   C   3 and 4 only
   D   2, 3 and 4 only
   (Paper 1/2)

5 Sound waves of velocity 340 m/s have a wavelength of 0.136 m. The frequency, in Hz, of the sound is
   A   0.0004
   B   46.24
   C   1250
   D   2500
   (Paper 2)

6 (a) The diagram shows a transverse wave.

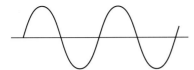

Copy the diagram and mark on it:

   (i) one wavelength, labelled λ

   (ii) the amplitude, labelled *a*.   *[2]*

(b) Give three examples of transverse waves. *[3]*

(c) Copy and complete the sentences:

   (i) In a transverse wave the vibrations are _____ to the direction of the wave motion.

   (ii) In a longitudinal wave the vibrations are _____ to the direction of the wave motion.   *[2]*

(d) Give one example of a longitudinal wave.   *[1]*

(Paper 3)

7 The diagrams show a ripple tank in which plane waves are approaching a barrier. There is a gap in the barrier.

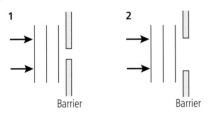

(a) Copy each diagram and draw three wavefronts to show the shape and spacing of the waves after they have passed through the gap in the barrier.   *[4]*

(b) Copy and complete the following sentences:

   (i) Sound waves are produced in air when an object _____.

   (ii) Longitudinal waves cannot be _____ .

   (iii) The _____ of a wave is the number of complete waves passing a point per second.   *[3]*

(Paper 3)

8 (a) Write down the equation linking the speed, frequency and wavelength of a wave.   *[2]*

(b) The diagram shows a ship using a sonar device to find the depth of the water by detecting the sea bed. Sound waves are transmitted from the ship and are reflected off the sea bed. They are detected back at the ship.

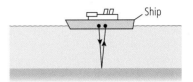

The speed of sound in sea water is 1500 m/s. The signal takes 1.2 s to be received at the ship after transmission.

   (i) Calculate the depth of the sea bed.

   (ii) The frequency of the sound waves used is 30 000 Hz. Calculate the wavelength of the sound waves.   *[5]*

(Paper 4)

8.1   Reflection of light

Supplement

LEARNING OUTCOMES

- Describe the formation of an image by a plane mirror
- Describe the properties of the image formed by a plane mirror
- Recall and use the law of reflection
- Explain the formation of the image by a plane mirror

Mirror images

If you have visited a 'Hall of Mirrors' at a funfair, you will know that the shape of a mirror affects what you see. To see an undistorted image of yourself, you need to look in a plane (i.e. flat) mirror.

The photograph shows the images of a toy in front of two plane mirrors. Whichever mirror you look at, you can see:

- the image in the mirror is the same size as the object
- the image is behind the mirror.

Which way would the images move if the toy was moved away from the mirror?

Images in a plane mirror

DID YOU KNOW?

Ambulances and police cars often carry a 'mirror image' sign at the front. This is so that a driver in a vehicle in front looking in the rear-view mirror can read the sign.

PRACTICAL

Investigating reflection of light by a plane mirror

To understand the nature of a mirror image, we need to understand how light behaves when it reflects from a mirror. Figure 8.1.1a shows how we can use a ray box and a mirror to investigate reflection. In the diagram, a narrow beam (or ray of light) reflects from the mirror. Alternatively, optical pins may be used as in Figure 8.1.1b.

- The straight line at right angles to the mirror is called the **normal**.
- The angle between the incident ray and the normal is called the **angle of incidence**, labelled *i* in the diagram.
- The angle between the reflected ray and the normal is called the **angle of reflection**, labelled *r* in the diagram.

If you make measurements of angle *r* for different measured angles *i* as described on the next page, you should find the two angles are always equal. Check this for yourself.

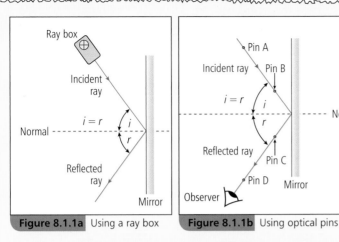

Figure 8.1.1a Using a ray box

Figure 8.1.1b Using optical pins

STUDY TIP

Remember that the angles *i* and *r* are always measured from the normal.

1 Mark a pencil line along the mirror and draw a line at right angles for the normal.

2 Mark the path of the incident ray at a measured angle of incidence to the normal.

3 Use the ray box to direct a ray (i.e. the incident ray) at the mirror or place optical pins A and B along the path of the incident ray. Measure and record the angle of incidence *i*.

4 Mark the path of the reflected ray either directly, if you are using a ray box, or observe the images of pins A and B such that A is directly behind B; then place two further pins C and D in line with the images of A and B.

5 Repeat the procedure for different angles of incidence.

The law of reflection

Measurements show that for any ray of light reflected by a mirror:

the angle of incidence, *i* = the angle of reflection, *r*

The light ray in Figure 8.1.1 gives the direction of the wavefronts in the light beam. To explain the law of reflection, remember that plane waves reflect from a straight barrier at the same angle as they move towards it. So the angle of the reflected waves and the incident waves to the reflecting barrier is the same. Therefore the reflected and incident rays are at the same angle to the normal.

Image formation by a plane mirror

Figure 8.1.2 shows how an image of a point object is formed by a plane mirror. The diagram shows the path of two rays of light from the object that both reflect off the mirror.

The angle of reflection of each ray is equal to its angle of incidence. The two reflected rays diverge as if they have come from the same point behind the mirror. This point is the image of the object. An observer looking along the two rays into the mirror sees an image of the object at this point.

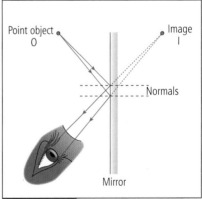

Figure 8.1.2 Image formation by a plane mirror

CONTINUED

LEARNING OUTCOMES

- Describe the formation of an image by a plane mirror
- Describe the properties of the image formed by a plane mirror
- Recall and use the law of reflection

- Explain the formation of the image by a plane mirror

PRACTICAL

Constructing a ray diagram.

The diagram in Figure 8.1.2 on the previous page is called a ray diagram. Construct your own ray diagram on a white sheet of paper to show the image is at the same distance behind the mirror as the object in front. Use a rule and a protractor. Make sure the angle of incidence is equal to the angle of reflection for each incident light ray.

Label the mirror, the normal, the object O and the image I on your diagram.

Measure the perpendicular distance from O and from I on the diagram to the mirror. The two distances should be equal.

Real and virtual images

The image seen in a mirror is a **virtual image**. When you look at a mirror image, the rays that reflect off the mirror into your eye appear to come from the image. A virtual image cannot be projected onto a screen like the movie images that you see at a cinema. Such an image is described as a **real** image because it is formed by focusing rays onto a screen.

STUDY TIP

The difference between a real image and a virtual image is an important distinction to understand.

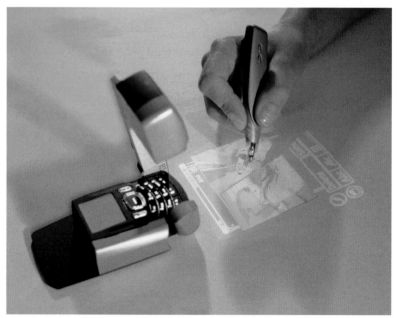

A real image

1 A point object O is placed in front of a plane mirror, as shown in Figure 8.1.3.

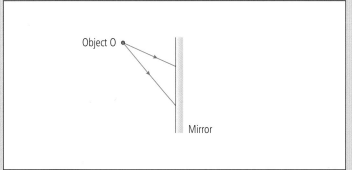

Figure 8.1.3

a Copy the diagram and complete the path of the two rays shown from O after they have reflected off the mirror.

b **i** Use the reflected rays to locate the image of O.

    ii Show that the image and the object are equidistant (i.e. the same distance) from the mirror.

2 a Two plane mirrors are placed perpendicular to each other.

    i Draw a ray diagram to show the path of a light ray at an angle of incidence of 60° that reflects off both mirrors.

    ii Measure the angle between the final reflected light ray and the initial incident ray.

b A woman stands 0.5 m in front of a plane mirror on a wall. How far is her image **i** behind the mirror **ii** away from her?

c In **b**, the woman is unable to see below her knees in the mirror. Use Figure 8.1.4 to explain why she cannot see her feet in the mirror.

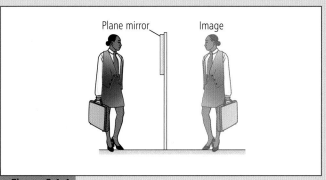

Figure 8.1.4

DID YOU KNOW?

You many have encountered the term 'normal' in mathematics, as well as the term 'tangent'.
A tangent is a straight line that touches a curve at a single point, and the normal at this point is perpendicular to the tangent

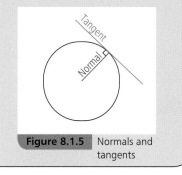

Figure 8.1.5   Normals and tangents

KEY POINTS

1 The normal at a point on a mirror is perpendicular to the mirror.

2 The image of an object seen in a plane mirror is virtual, the same size as the object, laterally inverted and at the same distance behind the mirror as the object is in front.

3 For a ray of light reflected by a mirror:

the angle of incidence = the angle of reflection

Refraction of light

LEARNING OUTCOMES

- Describe how to demonstrate refraction of light
- Describe the passage of a light ray through opposite sides of a rectangular transparent block

Refraction tests

When you have your eyes tested, the optician might test different lenses in front of each of your eyes. Each lens changes the direction of light passing through it. The change of direction is due to refraction which happens when light enters or leaves the lens. People with good eyesight don't need extra lenses. Each eye already has a perfectly good lens in it.

PRACTICAL

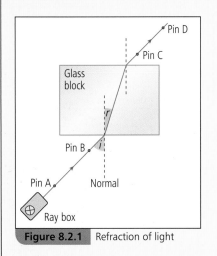

Figure 8.2.1 Refraction of light

Investigating refraction of light

1   Use a ray box and a rectangular glass block or optical pins as in Figure 8.2.1 to investigate refraction. If you use optical pins, pins C and D should be placed so they are in line with the images of pins A and B seen in line through the block.

2   Pins C and D should then be used to locate the point where the ray from A and B leaves the glass block. You should find that a light ray changes its direction at the boundary between air and glass unless it is along the normal.

Your investigation should show that a light ray directed along the normal passes straight through without being refracted. If the light ray is at a non-zero angle to the normal, it:

- bends towards the normal when it travels from air into glass. The angle of refraction, r, is smaller than the angle of incidence, i.
- bends away from the normal when it travels from glass into air. The angle of refraction, r, is greater than the angle of incidence, i.

Explaining refraction

Refraction is a property of all forms of waves including light and sound. Figure 7.3.2 in Topic 7.3 shows how we can demonstrate refraction using water waves in a ripple tank. The waves are slower in shallow water than in deep water. Because they change speed when they cross the boundary, they change direction:

- towards the normal when they cross from deep to shallow water and slow down
- away from the normal when they cross from shallow to deep water and speed up.

Light travels more slowly in glass than in air.

- When a light ray travels from air to glass, it refracts towards the normal because it slows down on entering the glass.
- When a light ray travels from glass to air, it refracts away from the normal because it speeds up on leaving the glass.

An eye test

Real and apparent depth

Don't jump into water unless you know how deep it is. Otherwise, you might find the water is much deeper than you thought. Light from the bottom refracts at the surface. This makes the water appear shallower than it really is. Figure 8.2.2 shows the refraction of a light ray from an object on the bottom of a swimming pool. The light ray bends away from the normal when it reaches the surface. The swimmer standing at the side sees a virtual image of the object above the object. To the swimmer, the apparent depth of the pool is less than the real depth.

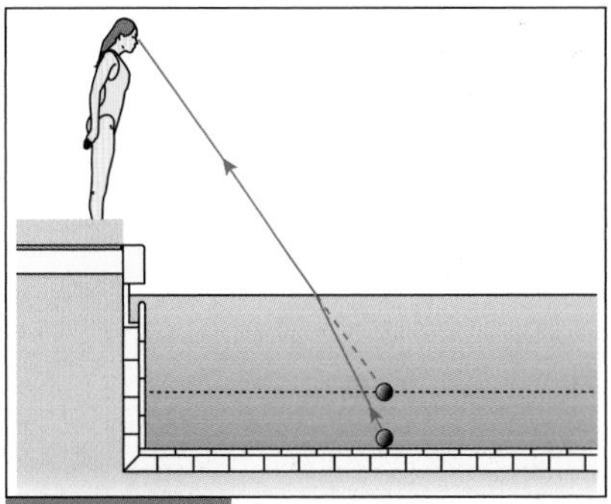

Figure 8.2.2 At a swimming pool

STUDY TIP

You must know which way the light will be refracted at the boundary between different materials.

PRACTICAL

Investigating refraction by a rectangular glass block

Test 1 Outline a rectangular glass block on a sheet of white paper. Use the arrangement in Figure 8.2.1 to observe a refracted light ray when it passes through opposite sides of the glass block. The light ray refracts towards the

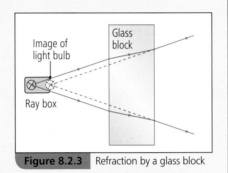

Figure 8.2.3 Refraction by a glass block

normal when it enters the block and it refracts away from the normal when it leaves the block.

Make measurements to find out if the direction of the light ray after leaving the block is the same as it was before it entered the block. You should find the direction is unchanged. This is because the opposite sides are parallel.

Test 2 Replace the slit plate of the ray box with a plate with two slits so the ray box produces two diverging rays. Direct the two rays into the side of the glass block as shown in Figure 8.2.3, so they emerge on the opposite side of the block. Look into the block along the outgoing rays to see an image of the ray box nearer to you. Trace the rays and locate the image.

KEY POINTS

1. Refraction of light is the change in direction of a ray of light when it crosses a boundary between two transparent substances including air.

2. Refraction is towards the normal when light travels from air to glass (or any other transparent substance).

3. Refraction is away from the normal when light travels from glass (or any other transparent substance) to air.

SUMMARY QUESTIONS

1. Copy and complete the sentences using words from the following list.

   away from   decreases   increases   towards

   a When a light ray travels from air into glass, its speed _____ and it bends _____ the normal.

   b When a light ray travels from glass into air, its speed _____ and it bends _____ the normal.

   c When a light ray travels from water into glass, it bends towards the normal because its speed _____.

2. In Figure 8.2.3:

   a Is the image real or virtual? Give a reason for your answer.

   b Explain why the image of the light bulb is nearer the glass block than the actual light bulb.

Supplement

Refractive index

A laser beam entering water

STUDY TIP

Supplement level candidates need to know and be able to use this equation in the examination.

Refractive index

When a ray of light travels from air into a transparent substance, its speed decreases. If the incident light ray is at a non-zero angle to the air-substance boundary, the light ray changes direction at the boundary. The change of direction is due to the change of speed at the boundary.

The speed of light in a transparent substance depends on the substance. For example, the speed of light in glass is about 12% faster than the speed in water. Therefore, the change of direction of a ray when it travels from air into a transparent substance depends on the substance.

The refractive index of a substance, *n*, is defined as:

$$\frac{\textbf{the speed of light in air}}{\textbf{the speed of light in the substance}}$$

For example,

- In air, light travels at a speed of 300 000 km/s, which means it travels a distance of 300 km in a thousandth of a second.

- In glass, light travels at a speed of 200 000 km/s, which means it would travel a distance of 200 km in glass in a thousandth of a second.

Therefore, the ratio $\dfrac{\text{speed of light in air}}{\text{speed of light in glass}} = \dfrac{300\,000\,\text{km/s}}{200\,000\,\text{km/s}} = 1.50$

So the refractive index of glass is 1.50.

The refractive index of a substance is a measure of the change of direction of a ray at non-normal incidence when it passes from air into the substance. For example, the refractive index of water is 1.33 and the refractive index of glass is 1.50. So glass refracts light more than water does.

PRACTICAL

Investigating how the angle of refraction varies with the angle of incidence

We can use a rectangular transparent block as shown in Figure 8.2.1.

The angle of refraction, *r*, is measured for different angles of incidence, *i*.

To make the measurements as accurately as possible:

1 Outline the block on a white sheet of paper and draw the normal at one of the longer flat sides of the rectangular block (whichever is used).

2 Use a protractor to draw straight lines at angles of 10°, 20°, 30° etc. to the normal. Label on the diagram as point P the point where the normal intersects the outline of the block.

3 Direct a ray along each straight line at P in turn. Label these lines 1, 2, 3, etc. corresponding to angles of incidence 10°, 20°, 30°, etc.

4 Mark on the outline of the block where the refracted ray in each case leaves the block for each angle of incidence. Label these points 1, 2, 3, etc. corresponding to angles of incidence 10°, 20°, 30°, etc.

5 Remove the block and draw straight lines from P to each marked point where a ray left the block. In each case, measure the angle of refraction.

6 Record all your measurements in a table. Some typical results are shown below. The results depend on the substance in the block.

$i\,/\,°$	$r\,/\,°$	$\sin i$	$\sin r$	$\dfrac{\sin i}{\sin r}$
10.0	6.5	0.174	0.113	1.54
20.0	13.0	0.342	0.225	1.52
30.0	19.0			

The measurements can be used to show that $\dfrac{\sin i}{\sin r}$ always has the same value, regardless of the angle of incidence. This relationship was first discovered in 1618 and is known as Snell's law, after its discoverer. Calculate the mean value of $\sin i/\sin r$ for your own measurements. As explained below, this is the refractive index of the block you tested.

The law of refraction

In Topic 7.3, we saw in Figure 7.3.3 how the refraction of waves is explained. Applying the explanation to rays of light, it can be shown that:

$$\frac{\sin i}{\sin r} = \frac{\text{the speed of the incident light waves}}{\text{the speed of the refracted light waves}},$$

where angles i and r are, respectively, the angles of incidence and of refraction.

Therefore, for light travelling from air into a transparent substance:

$$\frac{\sin i}{\sin r} = \frac{\text{the speed of light in air}}{\text{the speed of light in the substance}}$$

Since the refractive index of the substance, n, is defined as:

$$\frac{\text{the speed of light in air}}{\text{the speed of light in the substance}}$$

$$\frac{\sin i}{\sin r} = n$$

Note: given values of i and n, we can calculate r by rearranging the above equation to give:

$$\sin r = \frac{\sin i}{n}$$

and substituting the known values in the rearranged equation to find r.

KEY POINTS

1 Refractive index, n, of a transparent substance =

$$\frac{\text{speed of light in air}}{\text{speed of light in the substance}}$$

2 For a ray of light travelling from air into a transparent substance,

$$\frac{\sin i}{\sin r} = n$$

STUDY TIP

It's important to be able to carry out these types of experiment with confidence and accuracy.

SUMMARY QUESTIONS

1 a In the table above, when $i = 30°$, $r = 19°$, calculate the value of the refractive index.

b Determine the mean value of refractive index given by the data in the table above.

c The speed of light in air is 300 000 km/s. Use the value of refractive index calculated in **b** to work out the speed of light in the glass block that gave the results in the table.

2 The refractive index of water is 1.33.

a A ray of light enters a flat water surface at an angle of incidence of 35°. Calculate the angle of refraction of the light ray.

b The speed of light in air is 300 000 km/s. Calculate the speed of light in water.

Supplement

LEARNING OUTCOMES

- State the meaning of critical angle
- Describe total internal reflection
- Describe the use of optical fibres

The critical angle

When a ray of light is refracted as it passes from glass to air, as shown at point P in figure 8.4.1, it bends away from the normal.

If the angle of incidence in the glass is gradually increased, the angle of refraction increases until the refracted ray emerges along the boundary, as shown in Figure 8.4.2. The angle of incidence at this position is referred to as the **critical angle,** labelled c in Figure 8.4.2.

If the angle of incidence is increased beyond the critical angle, the light ray is **totally internally reflected** at P, as shown in Figure 8.4.3. The angle of reflection at P, r, is equal to the angle of incidence, i, when total internal reflection occurs.

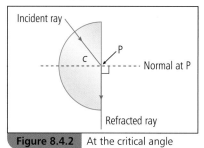

Figure 8.4.1 From glass to air	**Figure 8.4.2** At the critical angle	**Figure 8.4.3** Total internal reflection

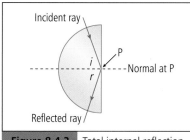

In Figures 8.4.1 and 8.4.2, because the rays travel from glass to air at point P,

$$\frac{\sin i}{\sin r} = \frac{\text{speed of light in transparent substance}}{\text{speed of light in air}} = \frac{1}{n}$$

as n is defined as the speed of light in air ÷ speed of light in the substance.

In Figure 8.4.2, the angle of incidence, i, = the critical angle, c, and the angle of refraction = 90°.

Applying these values to the above equation gives: $\frac{\sin c}{\sin 90°} = \frac{1}{n}$. Since $\sin 90° = 1$, then $\sin c = \frac{1}{n}$ or $\boldsymbol{n = \dfrac{1}{\sin c}}$

PRACTICAL

Investigating total internal reflection

Outline a semi-circular glass block on a sheet of white paper.

1 Use a ray box to direct a light ray at the centre of the flat side of the block (i.e. point P in Figure 8.4.1) through the curved side. Note that the light ray passes straight through the curved surface of the block without refraction. This is because, at the point where it enters the block, the light ray is directly along the normal at that point.

2 Adjust the angle of incidence of the light ray so the ray refracts into the air, as in Figure 8.4.1.

Note that some light does reflect **partially** at P back into the block but most of the light refracts. This partial internal reflection becomes total internal reflection when the angle of incidence exceeds the critical angle.

3 Increase the angle of incidence until the ray refracts along the boundary, as in Figure 8.4.2. Trace this ray into the block and measure the critical angle.

4 Increase the angle of incidence beyond the critical angle and observe that total internal reflection occurs, as in Figure 8.4.3. Make measurements to show that the angle of incidence is equal to the angle of reflection.

Optical fibres

Optical fibres are very thin glass fibres. We use them to transmit light or infra-red radiation. The rays can't escape from the fibre. Each ray entering a fibre at one end leaves the fibre at the other end even if the fibre bends around. This is because a ray in the fibre is totally internally reflected each time it reaches its boundary. Optical fibres are used in medicine to see inside the body without cutting the body open and in telecommunications to send signals securely.

- The medical endoscope is used by a surgeon to see inside a body cavity such as the stomach. The endoscope is inserted into the stomach via the patient's throat (see the photo). The endoscope contains two bundles of fibres, one to shine light into the cavity and the other to see the internal surfaces in the cavity. A tiny lens over the second bundle is used to form an image on the ends of the fibres in the bundle. The image can then be seen at the other end of the fibre bundle and viewed through a TV camera.

- A telecommunications optical fibre is used to carry digital signals in the form of light pulses. Such signals may be produced from an electrical signal by a 'light emitting diode' (or LED) in an electrical circuit. The pulses of light stay inside the fibre because any light reaching the fibre boundary is totally internally reflected. When the pulses reach the far end of the fibre, they are detected by a light sensor and converted into an electrical signal. The signals carried by an optical fibre can be detected only by the optical fibre receiver, unlike radio wave signals which can be detected by any receiver in the path of the radio waves. In addition, an optical fibre can carry much more information than a radio wave signal.

STUDY TIP

Make sure you understand the terms critical angle and total internal reflection.

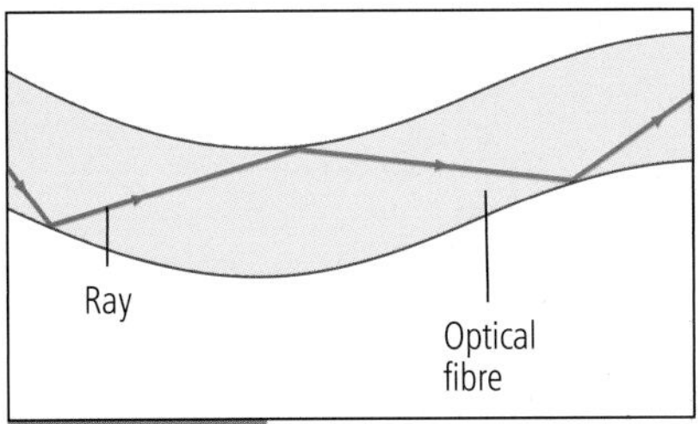

Figure 8.4.4 A ray of light in an optical fibre

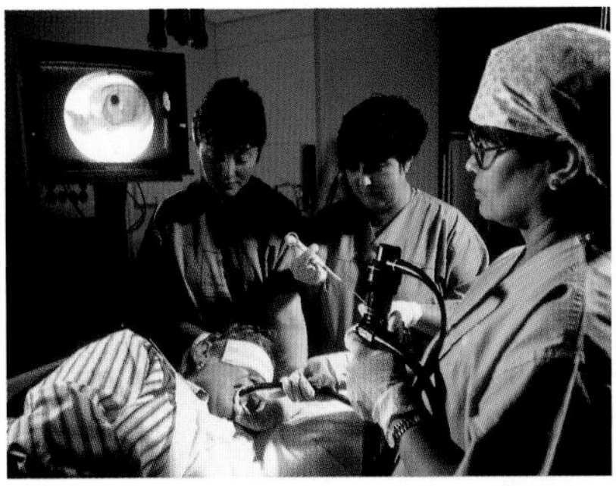

A stomach ulcer viewed through an endoscope

SUMMARY QUESTIONS

1 Copy and complete the following sentences using words from the list below.

**refraction   reflection   partial reflection
total internal reflection**

   a When a ray of light travels from air into glass and changes direction, it undergoes _____.
   b When a ray inside glass reaches the surface with air and stays in the glass, it undergoes _____.
   c When a ray passes from glass to air, some of the light undergoes _____.
   d When a ray inside glass reaches the surface with air, it undergoes _____ if the angle of incidence is less than the critical angle.

2 a Copy Figure 8.4.4 and show the path of an additional ray along the fibre.
   b State two advantages of using optical fibres instead of radio waves to carry signals.

KEY POINTS

When a light ray inside a transparent substance reaches the surface at an angle of incidence which is:

- less than the critical angle, it mostly refracts away from the normal (and some partial reflection occurs)
- equal to the critical angle, it refracts along the surface
- greater than the critical angle, it undergoes total internal reflection at the surface.

The converging lens

LEARNING OUTCOMES

- Describe how a thin converging lens works
- Recall the meaning of focal length
- Draw ray diagrams to show how an image is formed by a thin converging lens

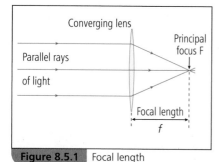

Figure 8.5.1 Focal length

STUDY TIP

You must be familiar with this experiment. You should know how to find the focused image and which way up it will be.

Converging lenses are used in optical devices such as the camera. Although a digital camera is very different from the very first cameras made over 160 years ago, they both contain a converging lens that is used to form an image.

A converging lens works by changing the direction of light passing through it. Each surface of the lens is a convex surface. The curved shape of the lens surface refracts the rays so they meet at a point. Figure 8.5.1 shows the effect of a converging lens on parallel rays.

- The point to where **parallel** rays directed straight at the lens are focused is the **principal focus F (or focal point)** of the lens.
- The distance from the lens to the principal focus is the **focal length, *f*,** of the lens.

PRACTICAL

Investigating the converging lens

1 Use the arrangement in Figure 8.5.2 to investigate the image formed by a converging lens.

2 With the object at different distances from the lens, adjust the position of the screen until you see a clear image of the object on it. The image is real because it is formed on the screen where the rays meet.

- When the object is a long distance away, the image is formed at the principal focus on the other side of the lens. This is because the rays from any point of the object are effectively parallel to each other when they reach the lens.

- If the object is moved nearer the lens, the screen must be moved further from the lens to see a clear image. The nearer the object is to the lens, the larger the image is. However, if the object is moved too near the lens, an image cannot be formed on the screen because the rays from the lens do not converge.

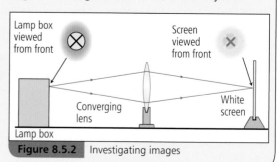

Figure 8.5.2 Investigating images

Ray diagrams

The position and nature of the image formed by a lens depends on:

- the focal length of the lens, and
- the distance from the object to the lens.

If we know the focal length and the object distance, we can draw a ray diagram to find the position and nature of the image. The ray diagram in Figure 8.5.2 shows how a converging lens forms a real image of an object. The object must be beyond the principal focus F of the lens. The image is formed beyond the principal focus on the other side of the lens.

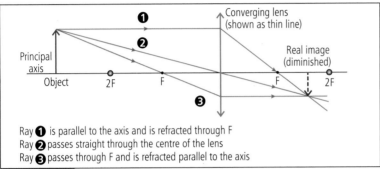

Ray ❶ is parallel to the axis and is refracted through F
Ray ❷ passes straight through the centre of the lens
Ray ❸ passes through F and is refracted parallel to the axis

Figure 8.5.2 Formation of a real image by a converging lens

The diagram shows that:

- three key 'construction' rays from a single point of the object are used to locate the image,
- the image is real, inverted and smaller than the object.

Notice that:

1 Ray 1 is refracted through F, the focal point of the lens, because it is parallel to the lens axis before the lens.
2 Ray 2 passes through the lens at its centre without change of direction. This is because the lens surfaces are parallel to each other at the axis.
3 Ray 3 passes through F, the focal point of the lens, before the lens so it is refracted by the lens parallel to the axis.

The image is diminished compared with the object because the object distance is greater than 2F. This is how a **camera** is used.

KEY POINTS

1 The principal focus of a converging lens is the point where parallel rays directed straight at the lens converge.

2 The focal length of a converging lens is the distance from the lens to the principal focus of the lens.

3 A real image is formed by a converging lens if the object is further away than the principal focus of the lens.

SUMMARY QUESTIONS

1 Copy and complete the following sentences using words from the list below.

at   near   far   from

a A converging lens forms a real image of an object on a screen.
  i   If the object is far from the lens, the image is formed on the other side of the lens _____ its principal focus.
  ii If the object is near the lens, the image is formed on the other side of the lens _____ its principal focus.

b A camera is used to photograph an object.

  i   If the object is not far from the camera, the distance from the lens to the film needs to be adjusted so that the film is _____ the principal focus of the lens.
  ii If the object is far from the camera, the distance from the lens to the film needs to be adjusted so that the film is _____ the principal focus of the lens.

2 a Draw a ray diagram to show how a converging lens forms a real image of an object.

b State whether the image is **i** upright or inverted, **ii** magnified or diminished.

Applications of the converging lens

Supplement

LEARNING OUTCOMES

- Explain what is meant by magnification
- Describe how a camera works
- Draw a ray diagram to show how a magnifying glass works

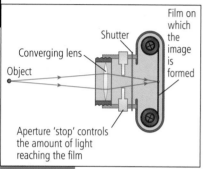

Figure 8.6.1 The camera

PRACTICAL

Investigating magnification

1 Use the arrangement in Figure 8.5.2 in Topic 8.5 to investigate how the size of the image depends on the distance from the lens to the screen. For different distances from the object to the lens:

- measure the image distance (i.e. the distance from the lens to the screen when the image is in focus on the screen)
- measure the diameter of the image.

2 What conclusion can you draw from your readings?

The image diameter and the object diameter are equal when the object and the image are equidistant from the lens. This is when the object is at 2F.

Magnification

The image formed by a converging lens may be smaller or larger than the object according to the distance from the object to the lens and the focal length of the lens. In general, we say that the image is:

- diminished if it is smaller than the object
- magnified if it is larger than the object.

If the object is further from the lens than the principal focus F of the lens, its image will always be real and inverted and on the other side of the lens. The image will be:

- diminished if the object is beyond 2F (i.e. beyond twice the distance from the lens to F)
- the same size as the object if the object is exactly at 2F
- magnified if the object is between F and 2F.

The camera

In a camera, a converging lens is used to produce a diminished, inverted, real image of an object on a film (or on an array of 'pixels' in the case of a digital camera). The position of the lens is adjusted to focus the image on the film, according to how far away the object is.

- For a distant object, the distance from the lens to the film must be equal to the focal length of the lens.
- The nearer an object is to the lens, the greater the distance from the lens to the film.

The projector

A projector lens produces a magnified real image of an object (such as a transparent slide) on a screen. Figure 8.6.2 shows the ray diagram for an object which gives a magnified, inverted, real image. The image can be seen in sharp focus on a screen placed where the image is located. If the screen is moved away from this position, the image becomes blurred and out-of-focus.

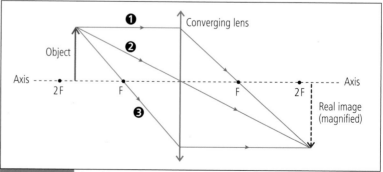

Figure 8.6.2 A real inverted magnified image

The magnifying glass

A magnifying glass is useful if you want to see fine detail on a small object such as a coin. A magnifying glass is a converging lens held close to the object. When the object is viewed through the magnifying glass, a magnified image of the object is seen. The photograph shows the image of a frond (divided leaf) seen through a magnifying glass. Notice that the image is upright (i.e. the same way up as the object).

The image is a **virtual image** (like an image seen in a mirror) because it cannot be formed on screen like a real image can. The ray diagram in Figure 8.6.3 shows how the virtual image is formed. The lens refracts the rays from the object so they appear to come from an image beyond the object. Provided the object is between the lens and its principal focus F, the image formed by the lens is always magnified, upright and virtual.

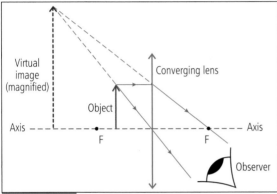

Figure 8.6.3 Ray diagram for a magnifying glass

PRACTICAL

Investigating the magnifying glass

Hold a converging lens over a printed page to observe a magnified virtual image of the print. Observe how the image is changed as the lens is moved away from the page. You should find the magnification increases until the image becomes too large and too blurred to see.

A converging lens as a magnifying glass

SUMMARY QUESTIONS

1 a Draw a ray diagram to show how a converging lens forms a magnified real image of an object.

b State whether the image is upright or inverted.

c Describe an application of the lens used in this way.

2 a Draw a ray diagram to show how a converging lens forms a magnified virtual image of an object.

b State whether the image is upright or inverted.

c Describe an application of the lens used in this way.

KEY POINTS

object–lens distance	image–lens distance	image description	application
beyond 2F	between F and 2F	real, inverted, diminished	camera
at 2F	at 2F	real, inverted, same size	
between F and 2F	beyond 2F	real, inverted, magnified	projector
at F	at infinity		
between F and the lens	same side of the lens as the object	virtual, upright, magnified	magnifying glass

Electromagnetic waves

LEARNING OUTCOMES

- Describe the spectrum of white light and how to produce it using a prism
- Recall the parts of the electromagnetic spectrum
- Recognise that all electromagnetic waves travel at the same speed through space
- State the approximate value of the speed of electromagnetic waves in air

The white light spectrum

Light from ordinary lamps and from the Sun is called **white light**. This is because it has all the colours of the visible spectrum in it. You see the colours of the spectrum when you look at a rainbow. You can also see them if you use a glass prism to split a beam of white light, as shown in the photograph, into separate colours. Light of a single colour is called **monochromatic** light.

Each colour of light is refracted slightly differently. This is because the speed of light in glass depends on its colour. As a result, the refractive index of glass (= speed of light in air/speed of light in glass) depends on its colour. The result is that the beam of white light is split into separate colours by the prism. This occurs due to refraction where it enters and where it leaves the prism. This effect is called **dispersion**.

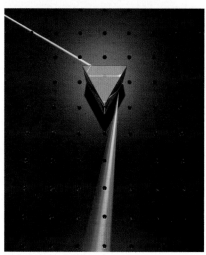

Dispersion of white light by a prism

STUDY TIP

The prism causes violet light to change direction more than red light.

PRACTICAL

Investigating the white light spectrum

1 Use a prism to split a narrow beam of white light from a ray box into the colours of the spectrum. Display the spectrum on a white screen. You should be able to see all the colours in the following order, as shown in Figure 8.7.1:

Red **O**range **Y**ellow **G**reen **B**lue **I**ndigo **V**iolet

(or Roy G Biv – the presence of indigo is sometimes disputed!)

2 Which colour is refracted most? The colour that is refracted most travels slowest in glass. You should find (as shown in the photograph) that violet is refracted most. So it travels slowest in glass. The speed of light in glass increases from violet to red across the spectrum.

3 Place a blackened thermometer on the screen just beyond each end of the spectrum. You should find a slight warming effect just beyond the red end of the spectrum. This is because the lamp produces **infra-red radiation** as well as light.

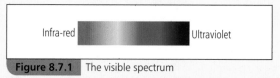

Infra-red   Ultraviolet

Figure 8.7.1 The visible spectrum

4 If a sensitive ultraviolet detector is placed just beyond the violet part of the spectrum, it should be possible to detect **ultraviolet radiation** there. This radiation does not have a heating effect like infra-red radiation, but it is harmful to the eyes.

The electromagnetic spectrum

Light, infra-red radiation and ultraviolet radiation are all part of the spectrum of electromagnetic waves. Figure 8.7.2 shows the electromagnetic spectrum. All electromagnetic waves travel through space at the same speed.

Notice the spectrum is continuous. The frequencies and the wavelengths at the boundaries between different parts are approximate as the different parts of the spectrum are not precisely defined.

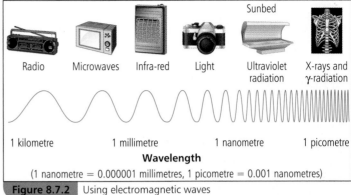

Figure 8.7.2 Using electromagnetic waves

(1 nanometre = 0.000001 millimetres, 1 picometre = 0.001 nanometres)

Electromagnetic waves are electric and magnetic disturbances that transfer energy from one place to another.

Supplement

- All electromagnetic waves travel through space at a speed of 300 000 km/s. Their speed through air is almost the same as their speed through space.
- Electromagnetic waves do not transfer matter. The energy they transfer depends on the **wavelength** of the waves. This is why waves of different wavelengths have different effects.

As explained in Topic 7.1, we can calculate the frequency from the wavelength (or the other way around) using the equation:

speed = frequency × wavelength

1 We can work out the wavelength if we know the frequency and the wave speed. To do this, we rearrange the equation into:

$$\text{wavelength (in metres)} = \frac{\text{wave speed (in m/s)}}{\text{frequency (in Hz)}}$$

2 We can work out the frequency if we know the wavelength and the wave speed. To do this, we rearrange the equation into:

$$\text{frequency (in Hz)} = \frac{\text{wave speed (in m/s)}}{\text{wavelength (in metres)}}$$

Note that 1 MHz = 1000 kHz = 1 000 000 Hz

> **STUDY TIP**
>
> LONG wavelength waves have a LOW frequency. SHORT wavelength waves have a HIGH frequency. Use the correct words!

KEY POINTS

1 Dispersion is the splitting of white light into the colours of the spectrum using a prism.

2 The main parts of the electromagnetic spectrum are, in order of decreasing wavelength:

radio waves

microwaves

infra-red radiation

light

ultraviolet radiation

X-rays

gamma rays

3 All electromagnetic waves travel at a speed of 300 000 km/s through space.

SUMMARY QUESTIONS

1 Copy and complete the sentences below using these words.

greater than   smaller than   the same as

a The wavelength of light waves is _____ the wavelength of radio waves.

b The speed of radio waves in a vacuum is _____ the speed of gamma rays.

c The frequency of X-rays is _____ the frequency of infra-red radiation.

2 a Copy and complete the electromagnetic spectrum below.

radio; _____; infra-red; visible; _____; X-rays; _____.

b Work out **i** the wavelength of radio waves of frequency 600 MHz, **ii** the frequency of microwaves of wavelength 0.30 m.

The speed of electromagnetic waves in a vacuum = 300 000 km/s.

Supplement

Applications of electromagnetic waves

LEARNING OUTCOMES

- Describe the main properties and uses of electromagnetic waves
- Recognise the dangers of X-rays, gamma rays and ultraviolet radiation

Light is just a small part of the electromagnetic spectrum. Our eyes cannot detect the other parts. The table below summarises the main sources, detectors, and uses of the different parts of the electromagnetic spectrum. Notice that light covers a very narrow range of wavelengths from about 400 nm for blue light to about 650 nm for red light, where 1 nanometre (nm) = 1 millionth of 1 mm.

More about X-rays

X-rays pass through body tissue but they are absorbed by bones and thick metal plates. To make a **radiograph** or X-ray picture, X-rays from an X-ray tube are directed at the patient. A light-proof cassette containing a photographic film is placed on the other side of the patient.

- When the X-ray tube is switched on, X-rays from the tube pass through the patient's body and leave a 'shadow' image of the bones on the film.
- When the film is developed, the parts exposed to X-rays are darker than the other parts. So the bones appear lighter than the surrounding tissue which appears dark. The developed film shows a 'negative image' of the bones.

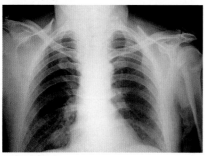

A chest X-ray

The electromagnetic spectrum

type of radiation	wavelength range	sources	detectors	main properties	uses
radio waves	1 km	radio, TV transmitters	receivers with aerials	reflected by metal sheets	radio and TV broadcasting
microwaves	1 m	microwave transmitters and ovens		reflected by metal sheets, absorbed by water	mobile phone communications, communications including satellite links, heating food
infra-red radiation	1 mm	hot objects	blackened thermometer bulb, infra-red camera	reflected by metal sheets, absorbed at or near the surface	infra-red cookers and heaters, infra-red cameras, intruder alarms, optical fibre communications, TV remote control
light	1 μm	glowing objects	eye, camera	reflected by mirrors, refracted by transparent substances	optical instruments, photography
ultraviolet radiation	1 nm	UV lamps, the Sun	fluorescent chemicals, UV film	makes fluorescent chemicals glow, absorbed by ozone layer	UV sun tan lamps, UV ink driers, security markers
X-rays	0.001 nm	X-ray tubes	film, Geiger tube (see Topic 15.1)	ionises substances, penetrates substances such as body tissue	X-radiography, security scans
gamma radiation		radioactive isotopes			Destruction of cancer cells, radioactive tracers, sterilising equipment, crack detection in metals

Safety matters

X-radiation or gamma radiation is dangerous and causes cancer. This is because X-rays and gamma rays create ions (charged atoms) in substances they pass through. High doses kill living cells and low doses cause cell mutation and cancerous growth. There is no evidence of a lower limit below which living cells would not be damaged. You will learn more about ionising radiation in Topic 15.1.

Anyone using equipment or substances that produce any form of ionising radiation must wear a **film badge** (see page 231).

Ultraviolet radiation is harmful to human eyes and can cause blindness. UV wavelengths are smaller than light wavelengths. UV rays carry more energy than rays of light. Too much UV radiation causes sunburn and can cause skin cancer.

• If you stay outdoors in hot weather, use skin creams to block UV radiation and prevent it reaching the skin.

• If you use a sunbed to get a suntan, don't exceed the recommended time. In addition, wear special 'goggles' to protect your eyes.

Radio waves and microwaves penetrate into the body and have a heating effect on body tissues (and any other substance that absorbs them). Infra-red radiation has a heating effect on skin tissue (as well as on the surface of any other substance that absorbs it).

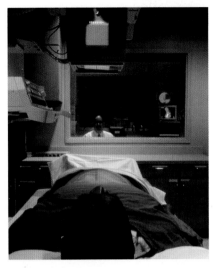
Taking a chest X-ray

STUDY TIP

Make sure you understand the process of ionisation. It is described in Topic 15.2.

SUMMARY QUESTIONS

1 Copy and complete the table below showing the type of electromagnetic radiation produced by each device **a** to **d**.

Device	infra-red	microwave	radio
a electric toaster			
b microwave oven			
c TV broadcast transmitter			
d remote handset			

2 Use words from the list to copy and complete the sentences below.

**infra-red radiation   gamma rays   light
microwaves   radio waves
ultraviolet radiation   white light   X-rays**

a _____ from the Sun is absorbed by the ozone layer.

b _____ includes all the colours of the spectrum.

c In a TV set, the aerial detects _____ and the screen emits _____.

d In a microwave oven, food absorbs _____, heats up and emits _____.

e _____ and _____ ionise living tissue.

PRACTICAL

An infra-red radiation test

Can infra-red radiation pass through paper? Use a TV remote handset to find out.

KEY POINTS

1 X-rays and gamma radiation damage living tissue when they pass through it.

2 X-rays are used in hospitals to take radiographs.

3 Ultraviolet radiation harms the skin and the eyes.

4 Light and infra-red radiation is used to carry signals in optical fibres.

5 Microwaves and radio waves are used for communications.

1 (a) Write down the law of reflection.

(b) Draw a labelled diagram showing a plane mirror, an incident ray, a normal where the incident ray hits the mirror, the reflected ray, the angle of incidence and the angle of reflection.

2 Which way is light bent (if at all) at the boundary between air and glass in the following circumstances:

(a) The light travels along a normal.

(b) The light hits the surface at an angle of incidence of 30° travelling from air to glass.

(c) The light hits the surface at an angle of incidence of 30° travelling from glass to air.

3 (a) Name the colours of the spectrum of white light in the correct order beginning with red.

(b) The diagram shows a ray of white light hitting a glass prism.

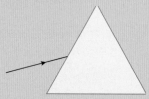

Copy and complete the diagram to show the dispersion of white light into the colours of the spectrum. Label the diagram so that it is clear which colour is refracted least and which colour is refracted most.

4 Suggest a source and a use for each of the following types of electromagnetic wave:

(a) microwaves

(b) infra-red

(c) ultraviolet

(d) X-rays.

5 Light enters a transparent rectangular block at an angle of incidence $I = 20°$. The angle of refraction is 13°.

(a) Calculate the refractive index of the material of the block.

(b) Draw a diagram to show the effect described.

6 Draw a ray diagram to show how a converging lens can be used as a magnifying glass.

1 The diagram shows a person looking at a mirror and viewing the image of a picture that is hanging on the wall behind him.

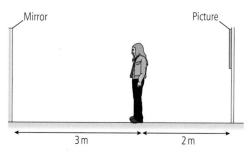

How far from the person is the image of the picture?

A 5 m   C 8 m

B 6 m   D 10 m

(Paper 1/2)

2 Which of the following is a true statement about refraction?

A When light travels from air to glass it refracts away from the normal because it slows down.

B When light travels from air to glass it refracts away from the normal because it speeds up.

C When light travels from air to glass it refracts towards the normal because it slows down.

D When light travels from air to glass it refracts towards the normal because it speeds up.

(Paper 1/2)

3 The diagram shows the electromagnetic spectrum.

γ-rays	x-rays				Microwaves

The missing labels are (1) infra-red, (2) radio, (3) ultraviolet and (4) visible.

From left to right on the diagram, the correct order for these labels is:

A 1, 2, 3, 4   C 3, 4, 1, 2

B 2, 1, 4, 3   D 4, 3, 2, 1

(Paper 1/2)

4 Which of the following is NOT a true statement about a converging lens?

A A converging lens is thicker in the middle than at the edge.

B A converging lens has a focus on each side.

Supplement

C A ray of light passes through the centre of a converging lens without changing direction.

D A converging lens forms an image on a screen that is the same way up as the object.

(Paper 1/2)

5 A ray of light enters a transparent, rectangular block at an angle of incidence $i = 30°$. The angle of refraction $r = 19°$.

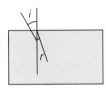

The refractive index n of the material of the block is

A 0.63

B 0.65

C 1.54

D 1.58

(Paper 2)

6 The diagrams show an incident ray directed through a semi-circular glass block to the centre of the flat surface.

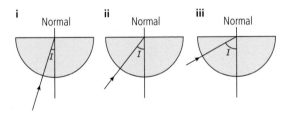

(a) Copy each diagram then draw the ray(s) that would be seen beyond the centre of the flat surface:

(i) where the angle of incidence is smaller than the critical angle

(ii) where the angle of incidence is equal to the critical angle

(iii) where the angle of incidence is greater than the critical angle. *[4]*

(b) Name the effect that you have shown in diagram **iii**. *[2]*

(Paper 3)

7 The diagram shows a glass block placed on a sheet of plain paper. A student has marked on the paper the passage of a ray of light through the block. The angle of incidence is 40°.

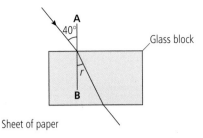

(a) (i) Name the line labelled AB.

(ii) Calculate the refractive index of the glass if the angle of refraction is 26°. Show your working.

(iii) The speed of light in air is 3×10^8 m/s. Calculate the speed of light in the glass. Show your working. *[6]*

(b) Explain why a swimming pool appears shallower than it is to an observer standing on the edge. You should copy the diagram and draw rays to illustrate your answer. *[3]*

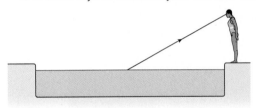

(Paper 4)

8 The ray diagram shows a converging lens with an object placed at a distance greater than the focal length from the lens.

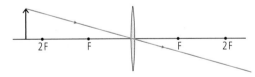

(a) (i) Copy and complete the ray diagram to find the position of the image. Label the position *I*.

(ii) State three properties of the image. *[5]*

(b) A small object is viewed through a converging lens. The lens has a focal length of 15 cm and the object is held 10 cm from the lens. The image seen is the right way up.

(i) State two other properties of the image.

(ii) State the name given to a lens used in this way. *[3]*

(Paper 4)

Sound waves

Supplement

LEARNING OUTCOMES

- Describe how sound waves are produced by vibrating sources
- Describe the longitudinal nature of sound waves
- Describe what compressions and rarefactions are

Making sound waves

Sound waves are easy to produce. Your vocal cords vibrate and produce sound waves every time you speak. Any object vibrating in air creates sound waves. You can use a loudspeaker to produce sound waves by passing alternating current through it. Figure 9.1.1 shows how to do this using a signal generator. This is an alternating current supply unit with a variable frequency dial. The frequency of the alternating current and hence the frequency of the sound waves can be changed by adjusting the frequency dial.

If you observe the loudspeaker closely, you can see it vibrating. It pushes the surrounding air backwards and forwards.

Making sound waves

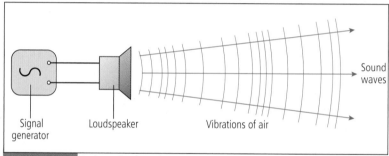

Figure 9.1.1 Using a loudspeaker

DID YOU KNOW?

When you blow a round whistle, you force a small ball inside the whistle to go around and around inside. Each time it goes around, its movement draws air in then pushes it out. Sound waves are produced as a result.

- If the frequency of the alternating current is decreased, the loudspeaker vibrates at a slower rate so it pushes the surrounding air backwards and forwards at a slower rate. The sound waves are therefore created with a lower frequency.

- If the frequency of the alternating current is increased, the loudspeaker vibrates faster so it pushes the surrounding air backwards and forwards at a faster rate. The sound waves are therefore created with a higher frequency.

Sound waves are longitudinal waves. When sound waves pass through air, the air molecules in the path of the sound waves move forwards and backwards along the direction in which the waves are moving. In other words, the molecules vibrate along the direction of motion of the waves. Sound waves are therefore longitudinal waves because the vibrations are along the direction they travel in.

Note This vibrating motion is much more significant than the individual random motion of the molecules just as the up-and-down motion of water molecules in sea waves is more significant than the random motion of the molecules.

STUDY TIP

You should know in what ways sound is similar to light and in what ways it is different.

Supplement

PRACTICAL

Making longitudinal waves

1 Use a slinky to demonstrate how sound waves travel. If you move one end backwards and forwards, longitudinal waves travel along the slinky.

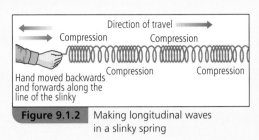

Direction of travel

Compression   Compression

Compression   Compression

Hand moved backwards and forwards along the line of the slinky

Figure 9.1.2 Making longitudinal waves in a slinky spring

2 Blow over the top of an empty bottle and see if you can make the bottle resonate with sound. Find out how the sound is affected by altering the water level in the bottle.

KEY POINTS

1 Sound waves are produced when a vibrating surface pushes and pulls on the surrounding substance.

2 Sound waves are longitudinal waves which means they vibrate along the direction in which the wave travels.

3 Sound waves in air consist of alternate compressions and rarefactions passing through the air.

The nature of sound waves

A loudspeaker sends sound waves through the air because its surface pushes and pulls repeatedly on the air. When the waves reach your ears, your eardrums vibrate so you hear sound.

As the surface vibrates, it moves forwards then backwards repeatedly. Each time it moves forwards into the surrounding air, it sends a **compression** wave through the surrounding air as a result of:

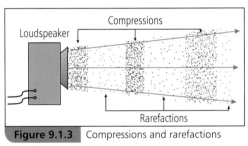

Compressions

Loudspeaker

Rarefactions

Figure 9.1.3 Compressions and rarefactions

• pushing the nearby air molecules forwards, so

• they bump into air molecules further away pushing them forwards, so they bump into air molecules further away pushing them forward …and so on.

Each time a vibrating surface moves backwards away from the surrounding air, it sends a **rarefaction** wave, the opposite of a compression wave, through the surrounding air as a result of:

• leaving a space which is then filled by nearby air molecules moving in, so

• they leave a space which is then filled by air molecules further away moving backwards,

• these air molecules leave space which is then filled by air molecules even further away moving in … and so on.

The result is that as the surface vibrates, it repeatedly sends a compression wave followed by a rarefaction wave into the surrounding air.

SUMMARY QUESTIONS

1 Copy and complete the following sentences using words from the list below.

**compressions
longitudinal waves
rarefactions   vibrations**

a Sound waves are caused by the _____ of an object.

b _____ are caused when a vibrating surface moves into the surrounding air.

c Sound waves are _____ because the _____ of the air molecules are along the direction in which the waves move.

d Sound waves consist of alternate _____ and _____ .

2 a Describe the motion of a point on a slinky coil when longitudinal waves pass along the slinky.

b Describe how the motion of air molecules is affected when sound passes through the air.

Supplement
Supplement
Supplement

LEARNING OUTCOMES

- Recognise that a medium is required to transmit sound waves
- Recall that sound waves cannot pass through a vacuum
- State the approximate range of frequencies that can be heard by the human ear

If you listen to a recording of your own voice, you may be surprised by the difference between your recorded voice and what you hear when you speak. The reason is that the sound of your own voice is carried to your ears by two routes:

- sound waves that pass from your larynx via your mouth into the air and then through the air to your ears, and
- sound waves that pass from your larynx to your ears directly through your head.

In comparison, your recorded voice is carried only by sound waves that pass through the air from the recorder to your ears. No sound waves pass directly through your head. Your recorded voice is the sound others hear when you speak.

Sound waves travel through solids, liquids and gases. You can test the passage of sound through a solid by placing a ticking watch in contact on a table and listening to it by placing your ear in contact with the table. The ticking of the watch can be clearly heard in this way even when it cannot be heard directly.

Sound waves cannot travel through a vacuum. This can be demonstrated by listening to an electric bell ringing in a 'bell jar', as shown in Figure 9.2.1.

- As the air is pumped out of the bell jar, the ringing sound fades away and can no longer be heard.
- When the air is let back into the bell jar, the ringing sound returns and can be heard again.

Without air in the bell jar, no sound waves can be created inside the jar.

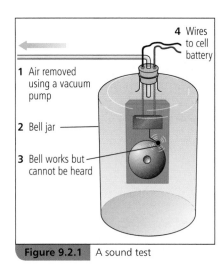

4 Wires to cell battery

1 Air removed using a vacuum pump

2 Bell jar

3 Bell works but cannot be heard

Figure 9.2.1 A sound test

STUDY TIP

Does this experiment show that light can travel through a vacuum? Yes, it does. You can still see the bell when the air is removed!

PRACTICAL

Hearing tests

1 Use a loudspeaker connected to a signal generator as in Topic 9.1 to find out the lowest and the highest frequency you can hear.

2 Work in a small group with each person in turn as the 'subject' at a fixed distance from the loudspeaker. For each subject, adjust the signal generator so the sound can be heard comfortably, then increase the frequency of the signal generator gradually until the subject can no longer hear the sound. This is the upper frequency limit of the subject's hearing. Repeat the test to find the lower frequency limit for the subject.

Young people can usually hear sound frequencies from about 20 Hz to about 20 000 Hz. Older people in general cannot hear frequencies at the higher end of this range. Sound waves above the upper limit of human hearing are referred to as **ultrasonic** waves or ultrasound.

Echoes

A sound echo is an example of reflection of sound. Echoes can be heard in a large hall or gallery which has bare smooth walls or in a cave with smooth walls. For example, if you clap your hands in a large hall, you should be able to hear an echo of the clap.

The echo is due to sound waves created when you clap your hands reflecting off the wall and returning to you. The further away you are from the wall, the longer the sound waves take to travel to the wall and back to you, so the longer the delay between the clap and the echo.

Unwanted echoes in buildings and halls can be eliminated by:

- covering the walls in soft fabric, which absorbs sound waves instead of reflecting them so no echoes will be heard
- making the wall surface uneven, not smooth, so reflected sound waves are scattered (i.e. 'broken up') so they cannot cause echoes.

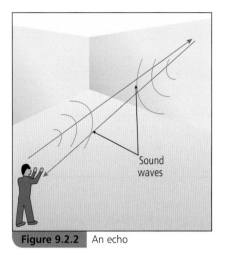

Figure 9.2.2 An echo

SUMMARY QUESTIONS

1 Copy and complete the sentences below using words from the list.

**absorbed   reflected   scattered
smooth   soft   rough**

a An echo is due to sound waves that are _____ from a _____ wall.

b When sound waves are directed at a _____ surface, they are broken up and _____.

c When sound waves are directed at a wall covered with a _____ material, they are _____ and not reflected.

2 a What is the highest frequency the human ear can hear?

b A sound meter is used to measure the loudness of sound. Describe how you would use the meter and the arrangement shown in Figure 9.2.3 to find out if high frequency sounds travel further than low frequency sounds. Assume the signal generator has a loudness control knob and a frequency dial that can be used to change the frequency of the sound from the loudspeaker.

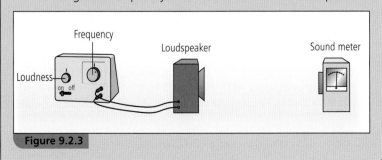

Figure 9.2.3

KEY POINTS

1 Sound waves can travel through liquids and gases and in solids.

2 Sound waves cannot travel in a vacuum.

3 An echo is due to sound waves that reflect from a smooth hard surface.

Supplement

LEARNING OUTCOMES

- Describe how to measure the speed of sound in air
- Explain how echo sounding can be used to measure distances
- State 'order of magnitude' values for the speed of sound in air, in liquids and in solids

A lightning stroke

DID YOU KNOW?

You can estimate how far away a lightning stroke is by counting the seconds between the lightning flash and the thunderclap clap then dividing by 3. The result gives you the approximate distance in kilometres. Divide by 5 for the answer in miles!

When you see a flash of lightning from a lightning stroke some distance away, you will hear the clap of thunder from the lightning stroke a short time later. The light waves from the lightning stroke reach you almost instantly, because light takes only about 3 millionths of a second to travel each kilometre. In comparison, sound waves in air take about 3 seconds to travel each kilometre.

PRACTICAL

Measuring the speed of sound in air

1 You need two people for this. You and a friend should stand on opposite sides of a field at a measured distance apart. You should be as far away from each other as possible but within sight of each other.

Figure 9.3.1 Measuring the speed of sound in air

2 If your friend bangs two cymbals together, you will see them crash together straightway but you will not hear them straightaway. This is because the 'bang' will be delayed as sound travels much slower than light. Use a stopwatch to time the interval between seeing the impact and hearing the sound.

Work out the speed of sound in air using the formula:

$$speed = \frac{distance}{time\ taken}$$

Note: If possible, swap positions so your friend creates the sound where you were at first and you detect the sound where your friend was at first. This is in case the wind affects the speed of sound in air. Use the two timing measurements to calculate the average time taken by the sound to travel the measured distance and then calculate the speed of sound from:

speed = distance/average time taken.

WORKED EXAMPLE

In an experiment, to measure the speed of sound in air, a student fired a starting pistol. Another student 600 m away measured the time between the flash and the bang at 1.8 s. Calculate the speed of sound in air.

Solution

$$\text{speed} = \frac{\text{distance}}{\text{time}} = \frac{600\,\text{m}}{1.8\,\text{s}} = 330\,\text{m/s}$$

The speed of sound in water

This was first measured in 1826 in Lake Geneva. A bell lowered under the water surface from a boat was struck with a hammer which also caused gunpowder to ignite with a flash. On seeing the flash, an observer in a boat 14 km away started a timer and stopped the timer when the sound was heard using an ear trumpet (an old-fashioned hearing aid) under the water. The speed was calculated using speed = distance/time. A timing of 9.8 s for a distance of 14 km gave a speed of 1430 m/s

Further investigation showed that the speed of sound in water also depends on the water temperature.

STUDY TIP

Sound waves travel over four times faster in water than in air.

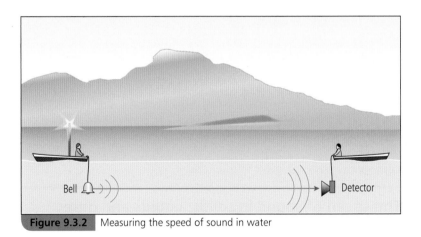

Figure 9.3.2   Measuring the speed of sound in water

Echo sounding

The depth of water beneath a boat or a ship can be measured by directing pulses of sound vertically downwards from a transmitter beneath the ship. Each pulse reflects from the sea bed and is detected by a detector such as a microphone at the same depth as the transmitter. Each detected pulse is an echo due to reflection from the sea bed (see Figure 9.3.3 on the next page).

CONTINUED

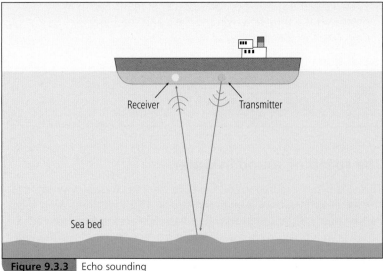

Figure 9.3.3 Echo sounding

STUDY TIP

Remember that the sound has to travel down to the sea bed and back up again to reach the receiver!

To find the depth, d, of the sea bed below the transmitter, the time taken, t, by each pulse to travel from the transmitter to the sea bed and back to the microphone is measured.

- In time t the distance travelled by the pulse $= vt$, where v is the speed of sound in water.
- The depth of the sea bed below the transmitter, $d = ½ \times$ the distance travelled by the pulse in time t.

Therefore $d = ½vt$

Note: The principle above is used to detect cracks inside metals. Ultrasonic waves are used because they are spread out less than sound waves at lower frequencies. Ultrasonic waves cannot be used for echo sounding because they are absorbed too easily compared with sound waves at lower frequencies.

The speed of sound in solids, liquids and gases

The speed of sound depends on the substance the sound is travelling through and on the temperature of the substance. The table on the opposite page shows the speed of sound in some different substances, all at the same temperature of 20 °C.

The table shows that in general, for the same temperature:

- sound travels much faster in solids and liquids than in gases,
- sound travels faster in solids than in liquids.

The table shows there are exceptions to these two general statements, for example sound travels faster in hydrogen gas than in liquid alcohol at the same temperature. However, the speed of sound in most gases is less than in most liquids at the same temperature.

Supplement

substance	physical state	speed of sound / m/s
air	gas	340
carbon dioxide	gas	270
hydrogen	gas	1320
alcohol	liquid	1210
water	liquid	1460
copper	solid	3560
iron	solid	5130
lead	solid	1230

STUDY TIP

You do not need to learn the numbers in the table but you do need to have an awareness of the relative speeds of sound in air, water and in solids.

SUMMARY QUESTIONS

Use speed of sound in air = 340 m/s.

1 a A clap of thunder from a lightning stroke was heard 6.0 s after the lightning flash was observed. Calculate the distance from the observer to where the lightning stroke occurred.

  b In an experiment to measure the speed of sound in air, a student blew a short blast on a whistle and raised an arm at the same time. An observer 500 m away heard the whistle blast 1.5 s after seeing the arm raised.

    i Explain why the whistle blast was heard by the observer after the arm was raised.

    ii Use the data to calculate the speed of sound in air.

2 In a test to measure the depth of the sea bed, sound pulses took 0.40 s to travel from the surface to the sea bed and back. Given that the speed of sound in sea water is 1350 m/s, calculate the depth of the sea bed.

3 A ship in fog near a cliff sounded its horn, as in Figure 9.3.4. An echo was heard 5.0 s later.

    i Calculate the distance from the ship to the cliff at the time when the horn was sounded.

    ii The ship's captain repeated the test a minute later and found the echo was heard 5.6 s later. Explain without calculation whether or not this and the previous measurement mean that the ship is moving away from the cliff.

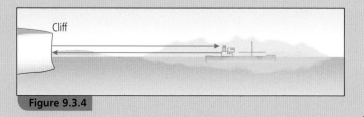

Cliff

Figure 9.3.4

KEY POINTS

1 An echo is a reflected sound.

2 The speed of sound in air is about 340 m/s.

3 The speed of sound in water is about 1400 m/s.

4 The speed of sound in many solids is more than 3000 m/s.

LEARNING OUTCOMES

- Relate the loudness and pitch of a sound to the amplitude and frequency of the sound waves
- Compare the loudness and pitch of different sounds from their waveforms

When you listen to music you usually hear sounds that are produced by instruments designed for the purpose. Even your voice is produced by a biological organ that has the job of producing sound.

- Musical notes are easy to listen to because they are rhythmic. The sound waves change smoothly and the wave pattern repeats itself regularly.
- Noise consists of sound waves that vary randomly in frequency without any pattern.

Making music

PRACTICAL

Investigating different sounds

Use a microphone connected to an oscilloscope to display the waveforms of different sounds.

Figure 9.4.1 Investigating different sound waves

1 Test a tuning fork to see the waveform of a sound of constant frequency.

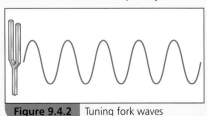

Figure 9.4.2 Tuning fork waves

2 Compare the 'pure' waveform of a tuning fork with the sound you produce when you talk or sing or when you whistle. You may be able to produce a pure waveform when you whistle or sing but not when you talk.

3 Use a signal generator connected to a loudspeaker to produce sound waves. The waveform on the oscilloscope screen should be a pure waveform.

Your investigations should show you that

- increasing the loudness of a sound makes the waves on the screen taller. This is because the **amplitude** of the sound waves (the maximum disturbance) is bigger the louder the sound is
- increasing **the frequency** of a sound (the number of waves per second) increases its pitch. This makes more waves appear on the screen.

Comparing waveforms

Figure 9.4.3 shows the waveforms for three different sounds A, B and C from the loudspeaker.

- A and B have the same amplitude and therefore the same loudness. However, C has a smaller amplitude so C is not as loud as A or B.
- C has more waves shown than A, which has more waves that B. This means C has a higher frequency (i.e. pitch) than A, which has a higher frequency than B.

Musical instruments

When you play a musical instrument, you create sound waves by making the instrument and the air inside it vibrate. Each new cycle of vibrations makes the vibrations stronger at certain frequencies. We say the instrument **resonates** at these frequencies. Because the instrument and the air inside it vibrate strongly at these frequencies when it is played, we hear characteristic notes of sound from the instrument.

A wind instrument, such as a flute, is designed so that the air inside resonates with sound when it is played. You can make the air in an empty bottle resonate by blowing across the top gently.

A string instrument, such as a guitar, produces sound when the strings vibrate. The vibrating strings make the surfaces of the instrument vibrate and produce sound waves in the air. The pitch of a vibrating string can be increased by shortening or tightening the string.

A percussion instrument, such as a drum, vibrates and produces sound when it is struck.

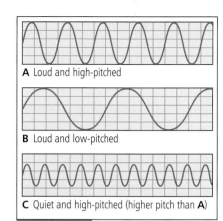

A Loud and high-pitched

B Loud and low-pitched

C Quiet and high-pitched (higher pitch than **A**)

Figure 9.4.3 Investigating sounds

PRACTICAL

A musical instrument

Using the apparatus in Figure 9.4.1 look at the waveform of a musical instrument. It changes smoothly, like the one in Figure 9.4.4. The waveform is a mixture of frequencies rather than a single frequency.

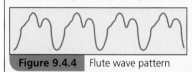

Figure 9.4.4 Flute wave pattern

KEY POINTS

1 The **loudness** of a note increases if the amplitude of the sound waves increases.

2 The **pitch** of a note increases if the frequency of the sound waves increases.

SUMMARY QUESTIONS

1 Copy and complete the following sentences using words from the list below:

decreases   increases   stays the same

a If the sound from a loudspeaker is made louder without changing its pitch, the amplitude of the sound waves _____ and the frequency of the waves _____.

b If the pitch of the sound from a loudspeaker is raised and the loudness is reduced, the amplitude of the sound waves _____ and the frequency of the waves _____.

c If a vibrating guitar string is shortened, the pitch of the sound it produces _____ and the frequency of the sound waves _____.

2 A microphone and an oscilloscope are used to investigate sound from a loudspeaker connected to a signal generator. What change would you expect to see on the oscilloscope screen if the sound is:

a made louder at the same frequency,

b made lower in frequency at the same loudness?

1 (a) Copy and complete the sentence.

   Sound waves are produced when objects _____ .

   (b) List four musical instruments and describe in each case what is vibrating to cause the sound.

2 Describe and explain the difference between transverse and longitudinal waves.

   Give one example of each type of wave.

3 (a) Describe an experiment to demonstrate that sound does not travel through a vacuum. Include a diagram in your answer.

   (b) How can you show with the same apparatus that light does travel through a vacuum?

4 (a) Explain what is meant by the term 'echo'.

   (b) A person is 30 m away from a reflecting surface. He claps his hands and hears the echo. If the speed of sound in air is 340 m/s, how long does it take from the time the person claps his hands to hearing the echo?

5 Describe an experiment to measure the speed of sound in air. Include a diagram in your answer and an explanation of how you would calculate the result from the readings.

6 Sound travels at different speeds in different substances.

   Which of the following gives the correct order for listing the speed of sound in different substances, from greatest to lowest speed?

   A   air, iron, water

   B   air, water, iron

   C   iron, water, air

1 Which of the following is NOT a true statement?

   A   Light is a transverse wave, sound is a longitudinal wave.

   B   Light waves can travel through a vacuum, sound waves cannot.

   C   Sound waves travel faster than light waves.

   D   Sound waves and light waves can be refracted.

   (Paper 1/2)

2 A spectator saw a cricketer hit the ball and heard the sound a short while later. The speed of sound in air is 340 m/s and the spectator was 80 m away from the batsman. What was the approximate time difference between the spectator seeing and hearing the ball being struck?

   A $\frac{1}{8}$ s   B $\frac{1}{4}$ s   C   1 s   D   4 s

   (Paper 1/2)

3 A sound wave is

   A   an electromagnetic wave

   B   a longitudinal wave

   C   a transverse wave

   D   a type of radio wave

   (Paper 1/2)

4 The loudness of a sound depends on

   A   the amplitude of the wave

   B   the frequency of the wave

   C   the speed of the wave

   D   the wavelength of the wave

   (Paper 1/2)

5 The pitch of a sound depends on

   A   the amplitude of the wave

   B   the frequency of the wave

   C   the speed of the wave

   D   the wavelength of the wave

   (Paper 1/2)

6 The diagram shows a 'slinky' spring. A longitudinal wave passes along the spring in the direction shown.

   Direction of travel ⟶

   (a) Show on a sketch the direction(s) of movement of the coils as the wave passes along the spring.   *[2]*

Supplement

(b) Copy and complete the sentences:

    (i) When the amplitude of a sound wave increases we detect this as an increase in _____ .

    (ii) When the frequency of a sound wave increases we detect this as an _____ in _____ . [3]

(c) The diagram shows the apparatus used to demonstrate an important property of sound waves.

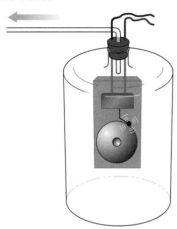

    (i) State the result that you would expect to notice when air is removed from the bell jar.

    (ii) State the result that you would expect to notice when air is allowed back in the bell jar.

    (iii) What do the results show about sound waves?

(Paper 3)

7 (a) (i) Explain how an echo is caused. You may draw a diagram to illustrate your answer.

    (ii) Explain briefly how the walls of a large room could be adapted to prevent unwanted echoes. [3]

(b) A ship is sailing in dense fog and it could be close to dangerous cliffs. The captain switches on the fog horn. The horn sounds and the echo is heard 4 s later. The speed of sound in air is 340 m/s.

    (i) Calculate the distance between the ship and the cliff.

    (ii) Next time the fog horn sounds it takes 3.5 s for the echo to be heard.

State whether the ship is further from, nearer to or at the same distance from the cliff. [3]

(Paper 3)

8 The diagram shows some cast-iron railings, 50 m long.

When the railings are hit with a hammer at A, a person standing at B hears the sound twice, through the air and through the railings.

The speed of sound in air is 340 m/s. The speed of sound in cast-iron is 5100 m/s.

    (a) Calculate the time taken for the sound to travel the 50 m in air.

    (b) Calculate the time taken for the sound to travel the 50 m through the railings.

    (c) Calculate the time between the two sounds being heard by the person at B. [5]

(Paper 4)

9 The diagrams of oscilloscope traces show the waveforms for three different sounds.

    (a) Compare the three traces in terms of the loudness and pitch of the sounds. [6]

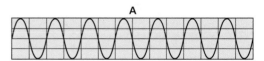

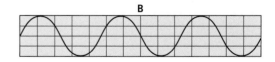

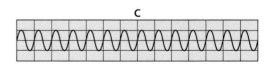

(b) State which property of the wave determines:

    (i) the loudness

    (ii) the pitch. [2]

(Paper 4)

10 Magnetism

10.1

Magnets

LEARNING OUTCOMES

- Explain why the two ends of a bar magnet are called the north-seeking pole and the south-seeking pole
- Recall that like poles repel and unlike poles attract
- Describe what is meant by magnetic induction

Using a compass

STUDY TIP

Some students confuse magnetic attraction and repulsion with positive and negative charges. Don't make that mistake!

Magnetic poles

A magnetic compass is a very useful device if you lose your way outdoors. The compass needle is a tiny 'bar' magnet that is pivoted at its centre of mass so it is free to turn horizontally.

When outdoors, the needle turns until one end points north and the other end points south. The effect is caused by the Earth's magnetic field. The end of the plotting compass that points north is called the 'north-seeking' pole and the other end the 'south-seeking' pole (usually referred to as the magnet's **north pole** (N) and **south pole** (S), respectively).

PRACTICAL

Investigating bar magnets

1 Suspend a bar magnet as shown in Figure 10.1.1 to check which end is its N-pole and which is its S-pole. Make sure the bar magnet is suspended horizontally and there are no other magnets nearby. If the ends are unmarked, label them N and S respectively.

2 Hold the N-pole of a second bar magnet near the suspended bar magnet as shown in Figure 10.1.2. You should find it attracts the S-pole of the suspended bar magnet and repels the N-pole.

3 Repeat the above test by holding the S-pole of the second bar magnet near the suspended bar magnet. This time, you should find it attracts the N-pole of the suspended bar magnet and repels its S-pole.

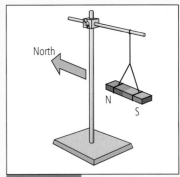

Figure 10.1.1 Checking the poles of a bar magnet

Figure 10.1.2 Like poles repel; unlike poles attract

The tests show the general rule that:

like poles repel; unlike poles attract

Magnetic induction

An unmagnetised iron bar placed near a bar magnet is always attracted to the bar magnet. For example, Figure 10.1.3 shows an unmagnetised bar XY placed end-on with a bar magnet a few centimetres away. Whichever way around the magnet is placed, there is always a force of attraction between XY and the bar magnet.

- In Figure 10.1.3a, the magnet's N-pole induces (i.e. brings about) an S-pole at the nearest end of XY (i.e. end Y). So end Y is attracted to the bar magnet.

- In Figure 10.1.3b, the bar magnet has been reversed so the magnet's S-pole is nearest to end Y. This induces an N-pole at end Y. So end Y is attracted to the S-pole of the bar magnet.

In other words, the magnet induces magnetism in XY such that the end of XY nearest to a pole of the bar magnet has the opposite polarity to that pole and is therefore attracted by the bar magnet.

When the bar magnet is removed, the magnetism induced in XY is lost.

Magnetic materials

Any iron or steel object can be magnetised (or demagnetised if it is already magnetised). Steel is referred to as a **ferrous** material because it contains iron. Any ferrous material can be magnetised or demagnetised. Only a few non-ferrous materials can be magnetised and demagnetised. Cobalt and nickel are two such examples. Oxides of these metals are used to make ceramic magnets and to coat magnetic tapes and discs.

STUDY TIP

Most metals are NOT magnetic.

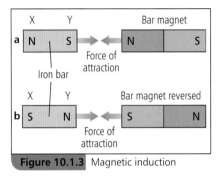

Figure 10.1.3 Magnetic induction

KEY POINTS

1 Like poles repel and unlike poles attract.

2 A ferrous material, such as steel, contains iron. Any ferrous material can be magnetised and demagnetised.

3 Magnetism is induced in a bar of ferrous material when a magnet is placed near it.

SUMMARY QUESTIONS

1 A bar magnet XY is freely suspended in a horizontal position so end X points north and end Y points south.
   a State the magnetic polarity of **i** end X, and **ii** end Y.
   b End P of a second bar magnet PQ placed near end X of bar magnet XY repels end X and attracts end Y. Explain this observation and state the magnetic polarity of end P.

2 A bar magnet is suspended horizontally as shown in Figure 10.1.1. When one end of an unmagnetised iron bar is placed near the N-pole of the bar magnet, the bar magnet is attracted to the iron bar.
   a Explain this observation.
   b State and explain what would be observed if the bar magnet was reversed.

DID YOU KNOW?

A magnetic compass in the northern hemisphere always points to a location referred to as the magnetic north pole. A magnetic compass in the southern hemisphere always points to the magnetic south pole. The location of the magnetic poles gradually changes, unlike the geographical poles which are fixed.

Magnetic fields

Supplement

LEARNING OUTCOMES

- Recall the pattern of magnetic field lines around a bar magnet
- Describe an experiment to plot the field lines of a magnetic field
- Describe how to magnetise and demagnetise a magnetic material

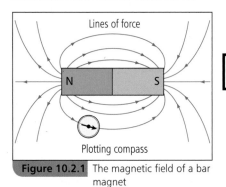

Figure 10.2.1 The magnetic field of a bar magnet

Lines of force

N   S

Plotting compass

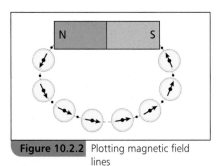

Figure 10.2.2 Plotting magnetic field lines

N   S

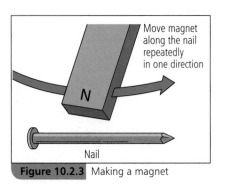

Move magnet along the nail repeatedly in one direction

N

Nail

Figure 10.2.3 Making a magnet

Magnetic field patterns

If a sheet of paper is placed over a bar magnet and iron filings are sprinkled onto the paper, the filings form a pattern of lines. The space around the magnet is called a **magnetic field.** Any other magnet placed in this space experiences a force due to the first magnet.

In Figure 10.2.1:

- iron filings form lines that end at or near the poles of the magnet. These lines are called **lines of force** (or sometimes 'magnetic field lines').
- a plotting compass placed in the field would align itself along a line of force, pointing in a direction away from the N-pole of the magnet and towards the magnet's S-pole. For this reason, the direction of a line of force is always from the north pole of the magnet to the south pole.
- magnetic forces are due to the interaction between magnetic fields.

PRACTICAL

Plotting a magnetic field

1   Draw an outline of a bar magnet on a sheet of white paper.

2   Mark a dot near the north pole of the bar magnet. Place the tail of the compass needle above the dot and mark a second dot at the tip of the compass needle. Move the compass so the tail of the needle is above the second dot and mark a new dot at the tip of the needle. Figure 10.2.2 shows the idea. Repeat the procedure until the compass reaches the S-pole of the magnet. Draw a line through the dots and mark the direction of the line from the N-pole to the S-pole of the bar magnet.

3   Repeat the procedure for further lines.

Making a magnet

An iron nail can be magnetised by moving one end of a bar magnet along the nail repeatedly in the same direction, as shown in Figure 10.2.3.

In Figure 10.2.3, the N-pole of a bar magnet is moved repeatedly along the nail from its head to its tip. When the nail is magnetised in this way:

- the head of the nail becomes an N-pole (as if it tries to repel the magnet as it approaches)
- the tip of the nail becomes an S-pole (as if it tries to attract the magnet it as it moves away).

The magnetic polarity of each end of the magnetised nail can be checked by holding each end of the nail in turn near a compass needle. If one end of the magnetised nail attracts the N-pole of the plotting compass and repels the S-pole, that end of the nail must be an S-pole (because unlike poles always attract).

To demagnetise the nail, the S-pole of the bar magnet (instead of the N-pole) can be moved repeatedly along the nail from head to tip. However, if the process is repeated too many times, the nail will be demagnetised then magnetised in the opposite direction.

Note: Another way to magnetise or demagnetise an iron (or steel) bar is to place it in a coil of wire which has a current passing through it. As explained in Topic 14.1, a magnetic field is produced in and around a coil when a current passes through it.

• To magnetise the bar, the current must be a steady direct current (e.g. supplied by a battery).

• To demagnetise the bar, the current must be an alternating current. See Topic 14.7 for more about alternating current. The bar is demagnetised by gradually removing it from the coil.

A magnet can also be

• magnetized by hammering it in a magnetic field

• demagnetized by hammering or heating it in the absence of a magnetic field

SUMMARY QUESTIONS

1 a Sketch the pattern of the magnetic field lines around a bar magnet. On your diagram, label the north and south poles of the bar magnet and indicate the direction of the field lines.

b Figure 10.2.4 shows a bar magnet XY and a plotting compass near end Y of the bar magnet. The needle of the plotting compass points towards end Y.

   i State the magnetic polarity of each end of the bar magnet.

   ii If the bar magnet was rotated gradually about its centre through 180°, describe and explain the effect on the direction of the plotting compass needle.

2 a Describe how you would use a bar magnet to magnetise an iron nail so its tip is an S-pole.

b The tip of an iron nail is held in turn near each end of a plotting compass needle. State whether the tip of the nail is an N-pole, an S-pole or is unmagnetised in each of the following possible observations.

   i The N-pole of the plotting compass is repelled by the tip of the nail and the S-pole is attracted by the tip.

   ii The N-pole of the plotting compass is attracted by the tip of the nail and the S-pole is repelled by the tip.

   iii The N-pole of the plotting compass is attracted by the tip of the nail and the S-pole is also attracted by the tip.

Supplement

PRACTICAL

Magnetise a nail

1 Use one end of a bar magnet as shown in Figure 10.2.3 to magnetise an unmagnetised nail. Use a plotting compass to determine the magnetic polarity of the nail.

2 Demagnetise the nail as described opposite. Use the plotting compass to check it has been demagnetised; in an unmagnetised state, both ends of the nail will be attracted to the plotting compass.

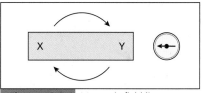

Figure 10.2.4 Magnetic field lines

KEY POINTS

1 A line of force (also called a field line) of a magnetic field is a line along which a plotting compass points.

2 The magnetic field lines of a bar magnet curve from the north pole of the bar magnet around to the south pole.

3 An iron or steel bar can be magnetised by:

• repeatedly moving a pole of a bar magnet along the bar in the same direction

• withdrawing the bar from a long coil carrying a direct current.

More about magnetic materials

Supplement

- Recall the meaning of 'hard' and 'soft' magnetic materials
- Distinguish between iron and steel in terms of their magnetic properties
- Explain why iron is used in an electromagnet and steel is used to make a permanent magnet

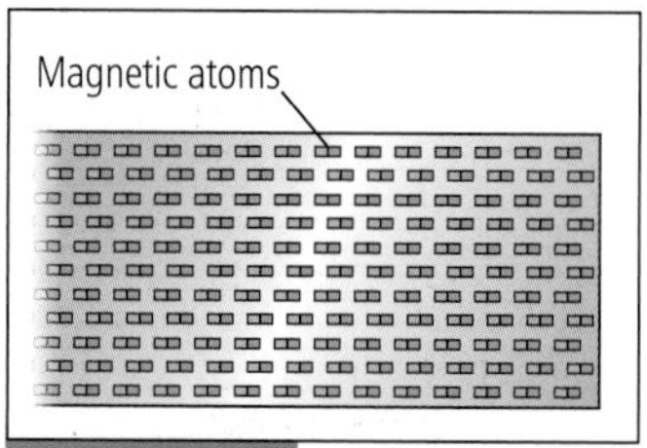

Magnetic atoms

Each atom of a magnetic material has its own magnetic field like that of a tiny bar magnet because of the arrangement and motion of the electrons in the atom. The atoms exert forces on each other due to their magnetism. In a magnet, the atoms are all lined up with each other, as shown in Figure 10.3.1, so the whole bar is magnetised. If the magnet is demagnetised, the overall pattern of alignment breaks up and there is no overall magnetisation.

Magnetic domains

In a magnetic material that is unmagnetised, the atoms align with each other in small regions called **domains**, as shown in Figure 10.3.2. The atoms in each domain are aligned with each other but the direction of alignment differs from one domain to another at random. As a result, the magnetic fields of the domains cancel each other out.

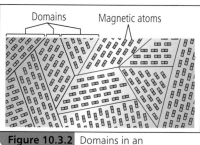

Figure 10.3.2 Domains in an unmagnetised iron bar

Note: Magnetic domains are not on the syllabus. They are introduced here to give a deeper understanding of magnetism.

Figure 10.3.1 A magnetised iron bar

STUDY TIP

Magnetism has always fascinated people and is a strange phenomenon in many ways. This information will help you to understand what it is.

Comparison of the magnetic properties of iron and steel

Steel is harder to magnetise and to demagnetise than iron, so it is referred to as a **hard** magnetic material. In comparison, iron is said to be a **soft** magnetic material because it is easy to magnetise and demagnetise.

- Steel rather than iron is used to make permanent magnets because iron loses its magnetism more easily; for example when it becomes hot or when it is struck repeatedly with a hammer. Hitting a magnetised iron bar repeatedly makes its atoms vibrate more, with the result that the atoms lose their overall alignment and so the bar loses its overall magnetism. Steel is much harder to demagnetise than iron because much more energy is needed to make the atoms in steel move out of their overall alignment.

- Iron rather than steel is used as the core of an electromagnet because iron is easier to magnetise and demagnetise than steel. As explained in Topic 10.2, an iron bar can be magnetised using a bar magnet or a current-carrying coil. In both cases, an external magnetic field is applied

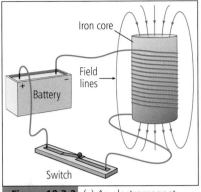

Figure 10.3.3 (a) An electromagnet   (b) Using an electromagnet

STUDY TIP

Although magnetism has been known about for many centuries it is still very much in use. There are many up-to-date uses (e.g. superconducting electromagnets in brain scanners) and research is still going on to discover more about magnetism.

to the iron bar. This field aligns all the atoms in the bar in the direction of the magnetic field. As a result, the whole bar becomes magnetised in the direction of the external magnetic field. An **electromagnet** consists of an iron 'core' with insulated wire coiled around it. When an electric current is passed through the wire, the core becomes magnetised. When the current is switched off, the core loses most of its magnetism. An electromagnet attached to a crane is used in scrap yards to lift car bodies or any other objects made of a ferrous material.

More about permanent magnets

Permanent magnets are made in different shapes including bars, flat rectangles, discs, cylinders, flat rings and U-shapes.

U-shaped magnets (sometimes referred to as horseshoe magnets) are bars bent into a U shape and then magnetised with opposite polarity at each end. The magnetic field in the gap between the two ends is very strong, as shown in Figure 10.3.4.

A loudspeaker magnet is a flat ring with a central cylinder of opposite polarity, as in Figure 10.3.5. The loudspeaker coil moves in the space between the flat ring and the central cylinder where the field is strongest. When an alternating current passes through the loudspeaker coil, the coil is forced to move in and out of the magnetic field as the current alternates. As a result, the diaphragm attached to the coil is made to vibrate, creating sound waves.

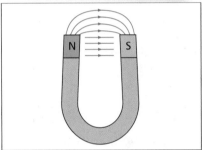

Figure 10.3.4 A horseshoe (U-shaped) magnet

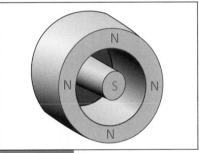

Figure 10.3.5 A loudspeaker magnet

SUMMARY QUESTIONS

1 Copy and complete the following sentences using words from the list below.

   hard   iron   keeps   loses   soft   steel

   a A permanent magnet is made from _____ which is magnetically _____ and therefore _____ its magnetism.

   b A _____ magnetic material such as _____ is easier to magnetise and demagnetise than a _____ magnetic material such as _____.

2 a Describe the essential features of an electromagnet.

   b Explain why the core of an electromagnet is made of iron rather than steel.

KEY POINTS

1 A 'hard' magnetic material is harder to magnetise and demagnetise than a 'soft' magnetic material.

2 In magnetic terms, steel is a hard and iron is soft.

3 Iron is used in an electromagnet and steel is used to make a permanent magnet.

SUMMARY QUESTIONS

1 Describe an experiment to plot the shape and direction of the magnetic field lines around a bar magnet.

2 Copy and complete the diagrams to show the shape and direction of the magnetic field between the magnetic poles.

(a)
N   S

(b)
S   N

(c)
N   N

(d)
S   S

3 (a) Describe and explain how to magnetise an iron nail using a magnet. Include a diagram in your answer.

(b) Describe how to demagnetise a magnetised iron bar.

4 Describe and explain the difference between a 'hard' magnetic material and a 'soft' magnetic material. Give an example of each.

5 (a) Describe what is meant by the term 'electromagnet'. Include a diagram in your answer.

(b) Give three examples of the uses of electromagnets.

PRACTICE QUESTIONS

1 A magnet will attract
   A aluminium
   B copper
   C iron
   D lead

(Paper 1/2)

2 The diagram shows plotting compasses around a bar magnet.

Which of the plotting compass needles are drawn pointing in the correct direction?
   A 1 and 2 only
   B 1 and 3 only
   C 1, 2 and 3 only
   D 4 only

(Paper 1/2)

3 The diagrams show the fields between two magnetic poles.

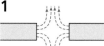

Which diagrams show two poles repelling each other?
   A 1 and 2 only
   B 1 and 3 only
   C 3 and 4 only
   D 4 only

(Paper 1/2)

Supplement

4 An unmagnetised iron bar placed near to a bar magnet is attracted to the bar magnet. This effect is called

A magnetic induction

B magnetic polarisation

C magnetic production

D magnetic repulsion

(Paper 1/2)

5 The shapes of the magnetic fields around two bar magnets of the same size are found using iron filings. The same amount of iron filings is used in each case. The magnetic field pattern of the stronger magnet will show

A lines of iron filings further apart

B lines of iron filings closer together

C most of the iron filings close to the centre of the magnet

D no difference between the two magnets

(Paper 1/2)

6 (a) (i) Draw a bar magnet as shown below, then sketch the pattern of field lines around it.

(ii) Show the direction of the field lines.

(iii) Name the apparatus that you would use to find the direction of the field lines. *[4]*

(b) If you are provided with an iron nail and a bar magnet, describe how you would magnetise the iron nail. You may include a diagram in your answer. *[3]*

(Paper 3)

7 You are provided with four metal bars of the same size labelled A, B, C and D. Pairs of bars are brought close to each other in turn. The table shows the observations.

bars	observation
A and B	attract
B and C	repel
A and C	attract
A and D	no effect
B and D	no effect
C and D	no effect

Each of the metal bars is either a magnet, a magnetic material or a non-magnetic material.

(a) State, with a reason for your answer, the magnetic property of each of the four metal bars A, B, C and D. *[6]*

(b) Describe, with a diagram, how you would use a plotting compass to find the shape and direction of the magnetic field around a bar magnet. *[3]*

(Paper 3)

8 Steel is described as a magnetically hard material. Iron is a magnetically soft material.

(a) Explain the terms **(i)** magnetically hard and **(ii)** magnetically soft. *[4]*

(b) (i) Suggest a use that is chosen for steel because it is magnetically hard.

(ii) Suggest a use that is chosen for iron because it is magnetically soft. *[2]*

(c) Describe in terms of the atoms why repeatedly hammering an iron magnet will cause the magnet to become demagnetised. *[3]*

(d) Draw the shape of the magnetic field between the poles of a horseshoe magnet. Show the direction of the field lines.

[3]

(Paper 4)

11 Electric charge

11.1   Static electricity

Supplement

LEARNING OUTCOMES

- Describe simple experiments to show how to produce and detect electric charge
- State that there are positive and negative charges
- State that unlike charges attract and like charges repel
- State that charge is measured in units called coulombs

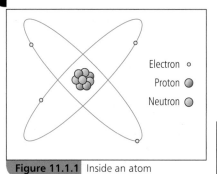

Figure 11.1.1   Inside an atom

STUDY TIP

You must learn the facts about the charge on protons, electrons and neutrons.

DID YOU KNOW?

If you take a woolly jumper off, listen out! You can hear it crackle as tiny sparks from static electricity are created. If the room is dark, you can even see the sparks.

Producing electric charge

To stick a balloon on a ceiling, all you need to do is to rub the balloon on your clothing before you touch it on the ceiling. The rubbing action gives the balloon **electric charge**. This stays on the balloon so we refer to it as **static electricity**. The charge on the balloon attracts it to the ceiling.

Inside the atom

To understand why certain materials can become charged, we need to consider what is inside an atom. Every atom has a positively charged nucleus composed of protons and neutrons. Electrons move about in the space around the nucleus.

- A proton has a positive charge.
- An electron has an equal negative charge.
- A neutron is uncharged.

An uncharged atom has equal numbers of electrons and protons. Only electrons can be transferred to or from an atom.

1. Adding electrons to an uncharged atom makes it negative (because the atom then has more electrons than protons).

2. Removing electrons from an uncharged atom makes it positive (because the atom has fewer electrons than protons).

Charge is measured in **coulombs**. We need over 6 million million million electrons to charge an object with 1 coulomb of charge. The magnitude of the charge of the electron is 1.6×10^{-19} coulombs. We will meet the coulomb again in Topic 11.4.

Charging by friction

Some insulators become charged by rubbing them with a dry cloth.

- Rubbing a perspex rod with a dry cloth transfers electrons from the surface atoms of the rod onto the cloth. So the perspex rod becomes positively charged.
- Rubbing a polythene rod with a dry cloth transfers electrons to the surface atoms of the rod from the cloth. So the polythene rod becomes negatively charged.

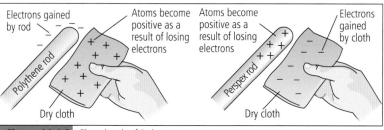

Figure 11.1.2   Charging by friction

PRACTICAL

The force between two charged objects

Two charged objects exert a force on each other. Figure 11.1.3 shows how you can investigate this force. Your results should show that:

- two objects with the same type of charge (i.e. like charges) repel each other

- two objects with opposite types of charge (i.e. unlike charges) attract each other.

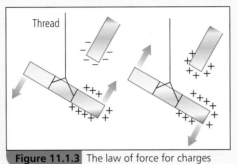

Figure 11.1.3 The law of force for charges

Like charges repel; unlike charges attract

STUDY TIP

Charged atoms are called _ions_. Atoms that gain electrons become negative ions. Atoms that lose electrons become positive ions. Any process that makes uncharged atoms into ions is called _ionisation_. You will meet ionisation in the next topic and in Chapter 15 on radioactivity.

The electroscope

Figure 11.1.4 shows an electroscope, a device that detects charge. We can use an electroscope to see if an object is charged. In Figure 11.1.4, the electroscope is charged by direct contact with a charged rod. The gold leaf of the electroscope is repelled by the metal plate when the electroscope is charged. This happens because they both gain the same type of charge.

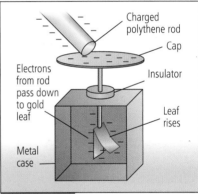

Figure 11.1.4 The electroscope

SUMMARY QUESTIONS

1 Choose words from the list to complete **a** and **b**:

to   from   loses   gains

a When a polythene rod is charged using a dry cloth, it becomes negative because it _____ electrons that transfer _____ it _____ the cloth.

b When a perspex rod is charged using a dry cloth, it becomes positive because it _____ electrons that transfer _____ it _____ the cloth.

2 a When rubbed with a dry cloth, perspex becomes positively charged. Polythene and ebonite become negatively charged. State whether or not attraction or repulsion occurs when:

i a perspex rod is held near a polythene rod

ii a perspex rod is held near an ebonite rod

iii a polythene rod is held near an ebonite rod.

b Glass is charged when it is rubbed with a cloth. When a charged glass rod is held near a positively charged rod, it is repelled by the rod.

i What type of charge, positive or negative, does a glass rod have when it is charged?

ii Does glass gain or lose electrons when it is charged?

KEY POINTS

1 Like charges repel; unlike charges attract.

2 Insulating materials that lose electrons when rubbed become positively charged.

3 Insulating materials that gain electrons when rubbed become negatively charged.

4 Charge is measured in coulombs.

Supplement

LEARNING OUTCOMES

- Describe what is meant by an electric field
- Explain what is meant by a line of force in an electric field
- Describe simple electric field patterns

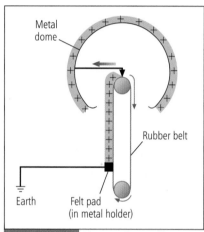

Figure 11.2.1 The Van de Graaff generator

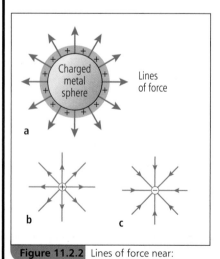

Figure 11.2.2 Lines of force near:
(a) a charged sphere
(b) a positive point charge
(c) a negative point charge

Lines of force

A **Van de Graaff generator** can make your hair stand on end. The dome charges up when the generator is switched on. Massive sparks are produced if the charge on the dome builds up too much.

The Van de Graaff generator charges up because:

- the belt rubs against a felt pad and becomes charged,
- the belt carries the charge onto an insulated metal dome,
- sparks are produced when the dome can no longer hold any more charge.

When the dome is charged, any other charged object brought near to it is repelled if it carries the same type of charge as the dome (or attracted if it carries the opposite type of charge). The charged dome is surrounded by an **electric field** due to its charge. Any other charged object near the dome is acted on (i.e. repelled or attracted) by the electric field of the dome.

A Van de Graaff generator in use

We can picture an electric field using the idea of **lines of force**. A line of force is the path a free positive charge would take if released in the field. Figure 11.2.2 shows the lines of force around a charged sphere and around a positive and a negative 'point charge'. Note the direction of the lines of force in each case (i.e. the direction which a free positive charge would move in the field). In all three examples, the electric field is said to be **radial**.

Electric field patterns

You can see the pattern of an electric field using the apparatus shown in Figure 11.2.3. Two conductors, connected to a Van de Graaff generator are submerged in castor oil sprinkled with semolina powder. The powder grains line up along the lines of force of the field.

- In Figure 11.2.4a, the field between the two oppositely charged parallel plates is said to be uniform because the lines of force are parallel to one another.
- In Figure 11.2.4b, one of the conductors has a curved section. The lines of force are concentrated at the curve because this is where the charge is most concentrated.

Note: the direction of the field lines is always from the positive conductor to the negative conductor.

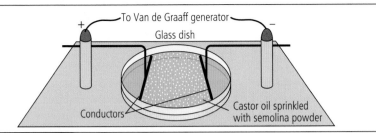

Figure 11.2.3 Demonstrating field patterns

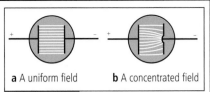

a A uniform field   **b** A concentrated field

Figure 11.2.4 Field patterns

Electric fields at work

The electrostatic paint sprayer is used to paint metal panels. The spray nozzle is connected to the positive terminal of an electrostatic generator. The negative terminal is connected to the metal panel. The panel attracts positively charged paint droplets from the spray.

The lightning conductor

A lightning strike is a massive flow of charge between a thundercloud and the ground. A lightning conductor on a tall building prevents lightning strikes by allowing the thundercloud to discharge gradually.

When a thundercloud is above a lightning conductor, a very strong electric field is created in the surrounding air. Air molecules near the conductor tip become **ionised** due to electrons being pulled off them. The ions then discharge the thundercloud so no lightning flash is produced. The conductor is joined to the ground by a thick copper strip. This allows charge to flow safely to earth.

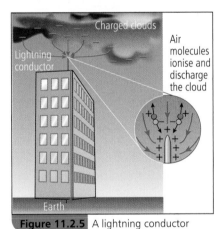

Charged clouds

Lightning conductor

Air molecules ionise and discharge the cloud

Earth

Figure 11.2.5 A lightning conductor

> **STUDY TIP**
>
> Make sure you understand the principles here so that you could explain other examples.

An electrostatic paint sprayer

> **KEY POINTS**
>
> 1 An electric field is the region around a charged object in which a second charged object experiences a force due to its charge.
>
> 2 A line of force is the path of a free positive charge released in an electric field.
>
> 3 The lines of force surrounding a point charge or a charged sphere are radial.
>
> 4 The lines of force between two oppositely charged parallel conductors are parallel to one another.

SUMMARY QUESTIONS

1 Copy and complete the following sentences using words from the list below.

   charge   field   force   ionisation

   a   The region around a charged metal sphere is referred to as an electric _____.

   b   The path of a charged particle released in this region is referred to as a line of _____.

   c   If a particle in this region does not experience a _____, its _____ must be zero.

2 Draw a diagram to show the pattern of the electric field:

   a   around a negative 'point charge',

   b   between two oppositely charged flat conductors.

   In both cases show the direction of the lines of force of the field.

Supplement

LEARNING OUTCOMES

- Distinguish between conductors and insulators
- Describe how an insulated conductor can be charged and discharged
- Explain the process of charging an insulated conductor by induction
- Explain conductors and insulators using the simple electron model

PRACTICAL

Testing conductors and insulators

Set up the circuit shown in Figure 11.3.1 and use it to find out which substances in the list below are insulators and which are conductors.

brass	cork
glass	lead
nylon	polythene
rubber	wood

A metal is a **conductor** of electricity. In Figure 11.3.1, the torch lamp lights up if a metal object is connected in the gap XY. The lamp will not light up if a plastic object is connected in the gap. Plastic is an **insulator**, not a conductor of electricity.

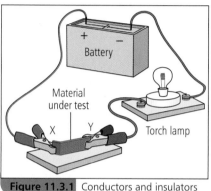

Figure 11.3.1 Conductors and insulators

Supplement

Metals conduct electricity because they contain **conduction electrons**. These electrons have broken free from atoms in the metal and they move about freely inside the metal. When a metal conducts electricity, electrons transfer electric charge because they are forced to move through the metal. Insulators cannot conduct electricity because all the electrons are held in atoms.

Charging and discharging a conductor

A conductor can hold charge only if it is insulated from the ground. If it isn't insulated, it will not hold any charge because electrons transfer between the conductor and the ground.

A conductor cannot be charged by rubbing it with a cloth because a cloth is not a good insulator. So it won't hold charge and neither will the conductor.

To charge an insulated conductor, we can bring it into contact with a charged object as shown in the charging of an electroscope in Figure 11.3.3.

- If the object is positively charged, electrons transfer from the conductor to the object. So the conductor becomes positive because it loses electrons.
- If the object is negatively charged, electrons transfer to the conductor from the object. So the conductor becomes negative because it gains electrons.

To discharge a charged conductor safely, a conducting path (e.g. a wire) needs to be provided between the object and the ground. The conducting path allows electrons to transfer between the conductor and the ground (Figure 11.3.2). We say the object is then **earthed.**

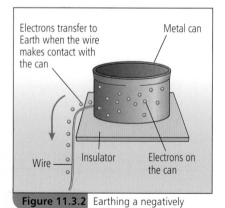

Electrons transfer to Earth when the wire makes contact with the can

Metal can

Wire — Insulator — Electrons on the can

Figure 11.3.2 Earthing a negatively charged conductor

Charging by induction

An insulated conductor can be charged from a charged body indirectly by **induction** as well as by direct contact. Figure 11.3.4 shows the process of charging an insulated metal sphere by induction using a negatively charged rod. Charging by induction gives the conductor the opposite type of charge to the charge on the rod. Also, no charge is gained or lost by the rod in this process.

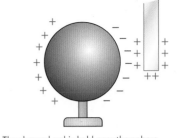

a The charged rod is held near the sphere

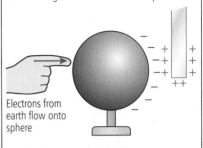

Electrons from earth flow onto sphere

b The sphere is earthed briefly

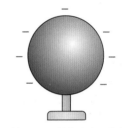

c The rod is removed. The sphere is left with an opposite charge to the rod

Figure 11.3.4 Charging by induction

PRACTICAL

Charge and discharge tests

1 Place a metal can on an electroscope as shown in Figure 11.3.3. Use a wire to make sure the can is discharged.

2 Charge a polythene rod by rubbing it with a dry cloth and place the rod in direct contact with the can. You should see that the electroscope leaf rises, indicating the presence of charge on the can and the electroscope.

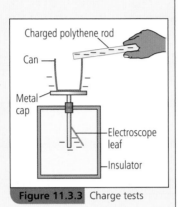

Charged polythene rod

Can

Metal cap

Electroscope leaf

Insulator

Figure 11.3.3 Charge tests

SUMMARY QUESTIONS

1 a Which of the following substances conduct electricity?

brass cork glass lead nylon polythene rubber wood

b i Explain why a metal object can hold charge only if it is insulated from the ground.

ii Describe how you would charge a metal sphere using a polythene rod.

2 a When a road tanker pumps oil or petrol into a storage tank, the connecting pipe must be earthed. If it is not, it could become charged. A build-up of charge would cause a spark which could make the fuel vapour explode.

i Why is the rubber hose of a petrol pump made of conducting rubber?

ii Why must the storage tank and the road tanker also be earthed before the fuel is pumped into the storage tank?

b In a dry climate, road vehicles carry a metal chain that is attached to the back of the vehicle so it touches the ground.

i Explain why the chain prevents the road vehicle from becoming charged?

ii Why is such a chain unnecessary in damp climates?

KEY POINTS

1 Any object that charge can pass through is a conductor.

2 Any object that charge cannot pass through is an insulator.

3 An insulated conductor can be charged by direct contact with a charged object.

4 An insulated conductor can also be charged by induction from a charged object.

Charge and current

Supplement

Supplement

LEARNING OUTCOMES

- State that an electric current is due to a flow of charge
- Recall the unit of electric current
- Describe the use of an ammeter

- State that charge is measured in coulombs
- Recognise that a current is a rate of flow of charge and use the equation $I = \dfrac{Q}{t}$

In Figure 11.4.1, each time the ball touches the left-hand plate, conduction electrons transfer to it from the plate so the ball becomes negatively charged. When the ball makes contact with the right-hand plate, electrons leave the ball so the ball becomes positively charged and is then attracted back to the left-hand plate. In this way, the ball transfers electrons, and therefore charge, across the gap.

The shuttling ball experiment

An electric current is a flow of charge. Figure 11.4.1 shows a demonstration in which a ball bounces 'to and fro' between two oppositely charged metal plates when the electrostatic generator is on. The ball is a table-tennis ball coated with conducting paint.

- When the ball touches the negative plate, it becomes negatively charged itself and is then repelled by the plate.
- The ball is then attracted onto the positive plate where it loses its negative charge and becomes positively charged.
- It is then repelled by the positive plate and attracted back to the negative plate and then repeats the 'to and fro' motion for as long as the generator is on.

The ball transfers charge from one plate to the other each time it moves across the gap. The meter in Figure 11.4.1 registers a tiny **electric current** because charge passes through it each time charge is transferred across the gap.

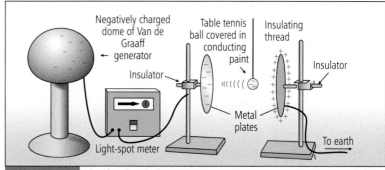

Figure 11.4.1 The shuttling ball experiment

If the plates are brought closer together, the ball shuttles between the plates at a faster rate. This makes the meter reading increase because the ball is 'ferrying' charge across the gap at a greater rate than before. So the electric current through the meter increases.

An electric current is due to a flow of electric charge.

Electric current

An **ammeter** is a meter designed to measure electric current. The unit we use to measure current is the **ampere (A)**. Figure 11.4.2 shows an electric circuit consisting of a torch lamp connected to a battery, a switch and an ammeter. We say the components are **in series** because the same amount of charge flows through each circuit component every second. The ammeter therefore measures the current in the torch lamp.

The direction of the current around an electric circuit is always shown on a circuit diagram from the + terminal of the battery to the − terminal, as in Figure 11.4.2.

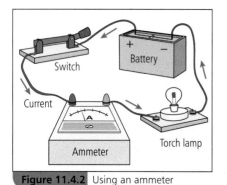

Figure 11.4.2 Using an ammeter

Measure the current in a torch lamp

1 Connect a torch lamp, a battery, a switch, an ammeter and a variable resistor in series with each other. Use the variable resistor to alter the brightness of the lamp.

2 Measure the lamp current when:

- the lamp is at normal brightness,
- the lamp can just be seen to be emitting light.

STUDY TIP

Do not confuse current with voltage.

Supplement

Electric charge is measured in coulombs (C). One coulomb is defined as the amount of charge passing a point in a circuit each second when the current is one ampere. In other words, a current of one ampere is equal to a rate of flow of charge of one coulomb per second.

1 **For a steady current *I* in a circuit**, the charge flowing *Q* in a certain time *t* is given by:

charge flowing = current × time
(in coulombs) (in amperes) (in seconds)

$$Q = It$$

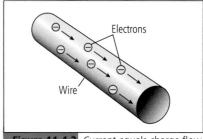

Figure 11.4.3 Current equals charge flow per second

2 **The direction in which electrons move around an electric circuit is from the − terminal of the battery to the + terminal.** This is opposite to the direction in which the current is marked on a circuit diagram. The reason is that the current direction convention was established as the direction of the flow of **positive charge** long before the discovery of electrons. Furthermore, an electric current can be due to the transfer of positive or negative charged particles. It just so happens that metals contain electrons that can move about freely inside the metal.

SUMMARY QUESTIONS

1 Copy and complete the following sentences using words from the list below:

**ampere ammeter charge
coulomb current stopwatch**

a The unit of current is the _____.

b To measure an electric current, we would need to use an _____.

c The unit of charge is the _____.

d To measure the _____ flowing along a wire carrying a constant _____, we would need to use an _____ and a _____.

2 a In the 'shuttling ball' experiment shown in Figure 11.4.1, describe and explain the effect on the ball of moving the plates further apart.

Supplement

b For each of the examples **i** to **iv** in the table below, calculate the missing quantity and give its unit in words.

current / A	3.5	ii	5	iv
charge / C	i	200	500	0.50
time / s	60	40	iii	20

STUDY TIP

The unit for current is often called the amp (short for ampere).

KEY POINTS

1 Electric current is due to a flow of charge.

2 Electric current is measured in amperes (A).

3 An ammeter is used to measure electric current.

Supplement

4 Current is the rate of flow of charge.

5 Electric charge is measured in coulombs (C).

6 charge flow = current × time (in seconds)

1 The diagram shows an atom. Name the parts labelled A, B and C.

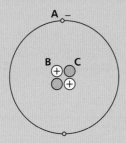

2 Copy and complete the following:

Rubbing a polythene strip with a dry cloth transfers _____ from the duster to the polythene. The polythene therefore becomes _____ charged, leaving the dry cloth _____ charged.

3 Describe a simple experiment using strips of insulating material hanging by insulating threads that will demonstrate the forces between charges. Describe the results that you would expect to observe and the conclusion that you would draw from the results.

4 The diagram shows a long-haired wig on top of the charged dome of a Van de Graaff generator.

Explain why the hairs 'stand on end'.

5 Write a list of four elements that are electrical conductors and four elements that are insulators.

6 (a) Write down the equation that links charge, current and time.

(b) State the units for each of the quantities in part **(a)**.

(c) A current of 0.8A is switched on. Calculate the amount of charge that flows in a time of 5 minutes.

Supplement

---

1 Study the list of materials:

aluminium copper cork iron
lead paper polystyrene rubber

The number of electrical conductors in the list is

A 3

B 4

C 5

D 6

(Paper 1/2)

2 Which of the following is NOT a true statement.

A A proton has a positive charge.

B An electron has a negative charge.

C A neutron has no charge.

D An atom has more protons than electrons.

(Paper 1/2)

3 When one or more electrons are removed from an atom, the atom is said to be

A electrolysed

B ionised

C magnetised

D polarised

(Paper 1/2)

4 The rate of flow of electric charge is called

A current

B power

C resistance

D voltage

(Paper 1/2)

5 The unit of electric charge is the

A amp

B coulomb

C ohm

D volt

(Paper 2)

6 The following diagram shows an electrostatic paint sprayer used to spray-paint a metal panel. The paint droplets are positively charged when they come out of the spray nozzle. The metal panel is charged negatively.

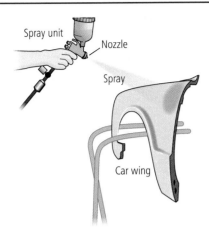

Spray unit
Nozzle
Spray
Car wing

(a) (i) Explain why the paint droplets spread out in the spray from the nozzle.

(ii) Explain why the metal panel is negatively charged.

(iii) Explain why this arrangement gives a more even coating of paint than using a paint brush. *[4]*

(b) The following diagram shows an electrostatic precipitator.

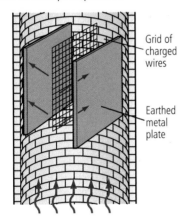

Grid of charged wires

Earthed metal plate

(i) What is the purpose of the grid of charged wires?

(ii) What is the purpose of the earthed metal plates?

(iii) What is the advantage to the environment of using an electrostatic precipitator? *[3]*

(Paper 3)

7 A student is supplied with some thin strips of material. Some are conductors, some are insulators. The student also has a power source, a lamp, a switch and some connecting wires.

(a) (i) Draw a circuit diagram using standard circuit symbols to show the circuit you would use to find which strips are conductors and which are insulators. The diagram is started for you.

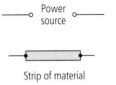

Power source

Strip of material

[3]

(ii) Name two conducting materials and two insulating materials. *[3]*

(b) An uncharged insulating rod is rubbed with an uncharged duster. The rod becomes negatively charged.

(i) State whether the duster, on rubbing the rod, will become positively charged, negatively charged or remain uncharged.

(ii) In terms of electrons, explain how the rod becomes negatively charged. *[3]*

(Paper 3)

8 The diagram shows a ball moving to and fro between oppositely charged metal plates.

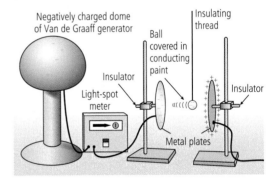

Negatively charged dome of Van de Graaff generator

Insulating thread

Ball covered in conducting paint

Insulator

Light-spot meter

Insulator

Metal plates

(a) (i) Explain carefully how the ball is caused to move to and fro. *[3]*

(ii) The meter shown in the diagram indicates a very small current whilst the ball moves to and fro. Define electric current in terms of electric charge. *[4]*

(b) A circuit is switched on. The steady current in the circuit is 0.6 A. Calculate the amount of charge that flows in 2 minutes. *[4]*

(Paper 4)

12 Electrical energy

12.1   Batteries and cells

Supplement

LEARNING OUTCOMES

- Recognise the significance of the value of the 'emf' of a battery
- Recall electromotive force (emf) is measured in volts
- Describe how to use a voltmeter to measure the emf of a battery or a pd in a circuit
- Explain the significance in terms of energy of the emf of a battery

Batteries in action

Calculators, digital watches, cameras, radios and cassette recorders all use batteries. They are used in cars, hearing aids, torches, toys and many other items. Batteries vary in size from the tiny batteries used in digital watches to heavy duty car batteries.

A battery consists of electric cells connected together. A cell (or a battery) is a source of electrical energy. Each cell transforms non-electrical energy into electrical energy. One terminal of a cell is positively charged and the other is negatively charged. For example:

Different types of battery

- A chemical cell transforms chemical energy into electrical energy. This happens because substances inside the cell react with each other. Figure 12.1.1 shows a simple non-rechargeable cell containing a zinc rod and a copper rod in dilute sulfuric acid. The zinc rod becomes negatively charged and the copper rod becomes positively charged as a result of the reactions in the cell.
- A solar cell transforms light energy into electrical energy. This happens because atoms in the solar cell release electrons as a result of absorbing light. This movement of electrons causes one terminal of the cell to become positively charged and the other negatively charged.

Electromotive force

When a cell is in a circuit, charge is forced to flow around the circuit by the cell. This flow of charge transfers energy from the cell to the circuit components then into the surroundings. In a chemical cell, this process continues until one of the substances in the cell has all reacted. In a solar cell, the process stops if there is no light reaching the cell.

The **electromotive force** (or 'emf') of a cell or a battery measured in **volts** is a measure of how much 'push' the cell or battery can provide to force charge around the circuit. The greater the emf, the more energy the cell can deliver for every electron that passes through it.

The emf of a cell, sometimes referred to as its 'voltage', depends on the substances in the cell. Batteries and cells are usually marked clearly in volts. For any given battery-operated device, the battery in it must be of the correct voltage. If the battery emf is too low, the device is unlikely to work. If the battery emf is too high, the device is likely to be damaged.

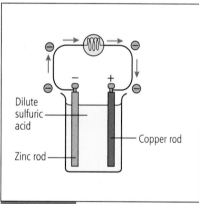

Figure 12.1.1 A simple chemical cell

Dilute sulfuric acid

Copper rod

Zinc rod

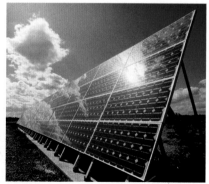

A panel of solar cells

The electromotive force (emf) of a source of electrical energy is the electrical energy it produces per coulomb of charge.

$$\text{emf of a battery in volts} = \frac{\text{electrical energy produced in joules}}{\text{charge in coulombs}}$$

The amount of electrical energy produced by a cell or battery depends on its emf. For every coulomb of charge that passes through it, a cell of emf 12 V produces 12 J of electrical energy. Therefore, a 12 V battery can deliver twice as much energy per coulomb of charge as a 6 V battery.

STUDY TIP

It is useful to remember that $1V = 1J/C$.

PRACTICAL

Measuring the emf of a battery

1 The emf of a cell or battery can be measured using a **voltmeter**, as shown in Figure 12.1.2.

2 The voltmeter must be connected so its + terminal is connected to the + terminal of the battery and its − terminal is connected to the − terminal of the battery.

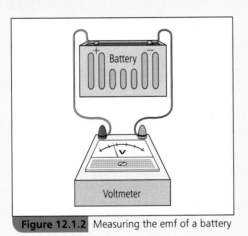

Figure 12.1.2 Measuring the emf of a battery

DID YOU KNOW?

Batteries marked at more than 12 V can give you a nasty shock if you touch a terminal. Mains electricity is lethal because it is supplied at a much higher voltage. Never ever touch a mains circuit or terminal.

SUMMARY QUESTIONS

1 Copy and complete the following sentences using words from the list below:

   an ammeter   charge   current   a voltmeter

   a The emf of a battery is measured using _____.

   b The _____ through a battery in a circuit is measured using _____.

   c The longer a battery is in a circuit, the greater the _____ that flows through it.

2 Describe how you would use a voltmeter to measure the emf of a battery.

KEY POINTS

1 Electromotive force (emf) is measured in volts.

2 The 'emf' of a battery is a measure of its 'push' or charge.

3 A voltmeter is used to measure the emf of a battery.

4 The emf of the battery is the electrical energy it produces per coulomb of charge.

12.2 Potential difference

Supplement

LEARNING OUTCOMES

- Explain what is meant by potential difference
- State that potential difference (pd) is measured in volts
- Describe how to use a voltmeter to measure the pd across a component in a circuit
- Explain the significance in terms of energy of the emf of a battery and a pd in a circuit

STUDY TIP

A voltmeter is connected IN PARALLEL with a component to measure the pd ACROSS the component.

DID YOU KNOW?

Figure 12.2.2 A digital meter

An ammeter or voltmeter with a pointer is called an **analogue** meter. An ammeter or a voltmeter with a digital read-out is called a **digital** meter. Whichever type of meter you use, always make sure it reads zero when it is not connected in a circuit or to a battery. To ensure an accurate reading of the analogue meter, the pointer must be viewed with the eye directly 'above' the pointer.

Energy transfer in a circuit

Figure 12.2.1 shows a lamp, a variable resistor and a battery connected in series. Charge from the battery has the **potential** to deliver energy to the circuit components. When it flows around the circuit, it transfers energy from the battery to the lamp and the variable resistor.

In Figure 12.2.1, a **voltmeter** is connected across the lamp. We say it is in **parallel** with the lamp. A voltmeter can be connected to any two points in a circuit. Its reading gives the **potential difference** (abbreviated to 'pd' and sometimes referred to as 'voltage') between those two points. This is a measure of the energy transferred by each electron as it passes between those two points.

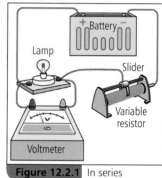

Figure 12.2.1 In series

- If a voltmeter is connected across the terminals of the battery, it measures the emf of the battery in volts, provided no energy is 'wasted' as heat inside the battery. The emf is a measure of the electrical energy produced in the battery for each electron that passes through it.
- If a voltmeter is connected across the lamp or the variable resistor, its reading in volts is a measure of the electrical energy supplied to the lamp or resistor by each electron.

Supplement

The potential difference (pd) across an electrical component is the electrical energy supplied to it per coulomb of charge flowing through it.

$$\text{potential difference (in volts)} = \frac{\text{electrical energy supplied (in joules)}}{\text{charge (in coulombs)}}$$

In Figure 12.2.1, in a given time:

the electrical energy produced by the battery = the electrical energy supplied to the lamp + the electrical energy supplied to the variable resistor

- The emf of the battery is the electrical energy per coulomb of charge produced by the battery.
- The potential difference across a component is the electric energy per coulomb of charge supplied to the component.

Therefore, from the above energy equation, we can say that:

emf of the battery = the pd across the lamp + the pd across the variable resistor

Note: The pd across the battery terminals is equal to the emf of the battery, provided no energy is wasted as heat inside the battery.

WORKED EXAMPLE

A 6 V lamp is connected in series with a 9 V battery and a variable resistor as shown in Figure 12.2.1. The variable resistor is adjusted until the lamp is at its normal brightness. The current through the lamp is then 2 A.

a Calculate:

  i the charge passing through the lamp in 10 s
  ii the electrical energy supplied to the lamp in this time
  iii the electrical energy supplied by the battery in this time.

b Explain why the pd across the variable resistor is 3 V.

Solution

a **i** charge = current × time = 2 A × 10 s = 20 C.

  ii electrical energy supplied to the lamp = pd × charge
    = 6 V × 20 C = 120 J

  iii electrical energy produced by
    battery = emf × charge = 9 V × 20 C = 180 J

b electrical energy supplied to variable resistor

  = energy produced by the battery − energy supplied to the lamp
  = 180 J − 120 J = 60 J

  pd across the variable resistor = $\dfrac{\text{energy supplied}}{\text{charge}} = \dfrac{60\,\text{J}}{20\,\text{C}} = 3\,\text{V}$

SUMMARY QUESTIONS

1 Copy and complete the following sentences using words from the list below:

charge   emf   energy   pd

  a _____ and _____ are measured in volts.

  b The bigger the _____ of a battery, the more _____ it can deliver for the same _____.

2 In a certain time interval, the battery in Figure 12.2.1 produces 120 J of electrical energy and delivers 80 J of this energy to the lamp.

  a Calculate the energy supplied to the variable resistor in this time.

  b The charge passing around the circuit in this time is 10 C. Calculate:

    i the emf of the battery,

    ii the pd across the lamp,

    iii the pd across the variable resistor.

  c The current during this time was 0.5 A. Calculate the duration of the time interval in seconds.

PRACTICAL

Measuring potential differences in a circuit

1 Connect a lamp, a variable resistor and a battery in series.

2 Use the slider of the variable resistor to adjust the brightness of the lamp so it is at less than normal brightness.

3 Use the voltmeter as shown in Figure 12.2.1 to measure:

  • the pd across the lamp.

  • the pd across the variable resistor; you will need to reconnect the voltmeter across the variable resistor to do this.

  • the emf of the battery; you will need to reconnect the voltmeter across the battery on its own to measure this.

4 Repeat the test for a different lamp brightness. Record your measurements.

  In both cases, assuming there is negligible resistance inside the battery, your measurements should show that:

  the emf of the battery = the pd across the lamp + the pd across the variable resistor

KEY POINTS

1 Electromotive force (emf) and potential difference (pd) are both measured in volts.

2 A voltmeter is used to measure a pd in a circuit.

3 The pd across a component is the electrical energy supplied to it per coulomb of charge.

LEARNING OUTCOMES

- Recall the definition of resistance and its unit
- Recognise that the current through a component depends on its resistance and on the pd across it
- Describe an experiment to measure the resistance of a wire
- Calculate resistance or pd or current for a component given two of the three quantities

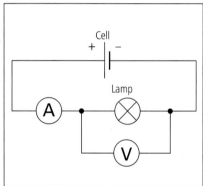

Figure 12.3.1 Using an ammeter and a voltmeter

WORKED EXAMPLE

The current in a wire is 2.0 A when the potential difference across it is 12 V. Calculate:

a the resistance of the wire

b the pd across the wire when the current in it is 5.0 A.

Solution

a $R = \dfrac{V}{I} = \dfrac{12}{2.0} = 6.0\,\Omega$

b $V = IR = 5.0\,\text{A} \times 6.0\,\Omega$
$= 30\,\text{V}$

Ammeters and voltmeters

The current in and the pd across a lamp or any other component can be measured at the same time using the circuit represented in Figure 12.3.1. The standard symbol for each component is used in the circuit diagram.

Look at the ammeter and the voltmeter in the circuit in Figure 12.3.1.

- The ammeter measures the current in the torch lamp. It is connected in **series** with the lamp so the current in them is the same. The ammeter reading gives the current in amperes (A) or milliamperes (mA) for small currents where 1 mA = 0.001 A.
- The voltmeter measures the potential difference across the torch lamp. It is connected in **parallel** with the torch lamp so it measures the pd across it. The voltmeter reading gives the pd in volts (V).

Electrons passing through a torch lamp have to push their way through lots of vibrating atoms. The atoms resist the passage of electrons through the torch lamp.

We define the **resistance** of an electrical component:

$$\text{resistance (in ohms)} = \frac{\text{potential difference (in volts)}}{\text{current (in amperes)}}$$

The unit of resistance is the **ohm**. The symbol for the ohm is the Greek letter Ω ('omega').

We can write the definition above as:

$$R = \frac{V}{I}$$

where V = potential difference in volts
I = current in amperes
R = resistance in ohms.

Note: Rearranging the above equation $R = \dfrac{V}{I}$ gives $V = IR$ or $I = \dfrac{V}{R}$

The last equation shows that the current in a component in a circuit can be increased if the pd across it is increased or the resistance of the component is decreased, for example as in a variable resistor.

PRACTICAL

Investigating the resistance of a wire

Does the resistance of a wire change when the current in it is changed? Figure 12.3.2a shows how we can use a variable resistor to change the current in a wire. Make your own measurements and use them to plot a current–potential difference graph like the one in Figure 12.3.2b.

current / A	0	0.05	0.10	0.15	0.20	0.25
potential difference / V	0	0.50	1.00	1.50	2.00	2.50

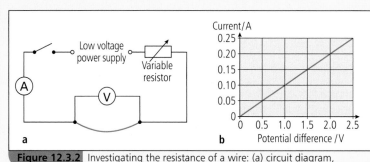

a Discuss how your measurements compare with the ones from the table used to plot the graph in Figure 12.3.2b.

b Calculate the resistance of the wire you tested and the resistance of the wire that gave the results in Figure 12.3.2b.

Figure 12.3.2 Investigating the resistance of a wire: (a) circuit diagram, (b) a current–potential difference graph for a wire

Current–potential difference graphs

For a wire, the graph in Figure 12.3.2b and your own graph should show:

- a straight line through the origin
- that the current is proportional to the potential difference.

Reversing the potential difference makes no difference to the shape of the line. The resistance is the same whichever direction the current is in.

The graph shows that the resistance (= potential difference/current) is constant. This is known as Ohm's law:

The current in a resistor at constant temperature is proportional to the potential difference across the resistor.

If a filament lamp is tested instead of a wire, the measurements give a graph which curves, as shown in Figure 12.3.3. This is because the resistance of the filament lamp increases as the current in it becomes hotter. This is because its atoms vibrate more, so they oppose the flow of electrons more.

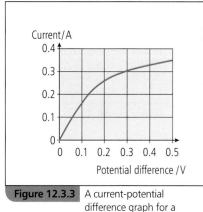

Figure 12.3.3 A current-potential difference graph for a filament lamp

Supplement

SUMMARY QUESTIONS

1 a Draw a circuit diagram to show how you would use an ammeter and a voltmeter to measure the current in and the potential difference across a wire.

b The potential difference across a resistor was 3.0 V when the current in it was 0.5 A. Calculate the resistance of the resistor.

2 Calculate the missing values in each line of the table.

resistor	current / A	potential difference / V	resistance / Ω
W	2.0	12.0	?
X	4.0	?	20
Y	?	6.0	3.0
Z	0.5	12.0	?

KEY POINTS

1 Resistance R (in ohms) = $\dfrac{\text{potential difference } V \text{ (in volts)}}{\text{current } I \text{ (in amperes)}}$

2 $R = \dfrac{V}{I}$ rearranged gives

$$V = IR \text{ or } I = \dfrac{V}{R}$$

3 The current in a resistor at constant temperature is proportional to the potential difference across the resistor.

Supplement

More about resistance

LEARNING OUTCOMES

- State that the resistance of a uniform wire increases if its length is increased
- State that the smaller the diameter of wire, the greater its resistance is

- Recognise that the resistance of a wire depends on the material it is made from and is:
  1 proportional to its length
  2 inversely proportional to its area of cross-section

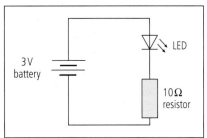

Figure 12.4.1 Using a resistor

Resistors

A **resistor** is a component designed to have a certain resistance. Resistors are usually made from metal wires or carbon. The symbol for a resistor is shown in the circuit diagram in Figure 12.4.1. Resistors are used in circuits to limit the current in the circuit.

The circuit diagram in Figure 12.4.1 shows a circuit with a 3.0 V battery in series with a light-emitting diode (LED) and a 10 Ω resistor. An LED emits light when a current passes through it. It is often used as an indicator in an electronic circuit, as explained in Topic 13.8. However, if too much current passes through it, it overheats and no longer conducts charge or emits light.

Because the pd across the resistor can never be more than the battery emf of 3.0 V, the current in the resistor can never exceed 0.3 A (= 3.0 V/10 Ω).

Wires and resistance

A wire that has the same diameter all along its length is said to be **uniform**. The resistance of a uniform wire depends on:

- its length – the longer the length of the wire, the greater its resistance
- its diameter – the smaller a wire's diameter, the greater its resistance
- the material it is made from – for example, a metal such as copper conducts more readily than a metal such as steel. So a copper wire of the same diameter and length as a steel wire has less resistance.

PRACTICAL

Make a wire-wound resistor

1 A wire-wound resistor consists of a wire wrapped around a suitable insulating tube. Constantan wire is often used because constantan, a metal alloy, does not rust or oxidise. Use the circuit in Figure 12.3.2 of Topic 12.3 to measure the resistance of different measured lengths of constantan wire (or a wire made of other suitable material) up to 1m in length. For each length,

- use the variable resistor to adjust the current to the same value and measure the pd across the wire for this current,

- use the equation 'resistance $= \dfrac{\text{pd}}{\text{current}}$

length of wire / m	0	0.20	0.41	0.60	0.79	1.00
current / A	1.5	1.5	1.5	1.5	1.5	1.5
pd / V	0	1.8	3.7	5.4	7.2	8.9
resistance / Ω	0	1.2	2.5	3.6	4.8	5.9

2 Record your measurements and the results of your calculations in a table as shown above.

3 Use your results to plot a graph of resistance against length as shown in Figure 12.4.2.

The length of wire needed to make a resistor of any value less than 6 Ω can then be found from the graph. For example, we can see from Figure 12.4.2 that a length of 0.50 m of the wire tested would be needed to make a resistor of resistance 3.0 Ω. From your graph, find the length of wire needed to make a resistance of 3.0 Ω. It will not necessarily be the same as that given by Figure 12.4.2 because your wire might have a different diameter and/or be made of a different material. Because the line on the graph is straight throught the origin, the graph shows that the resistance of the wire is directly proportional to its length.

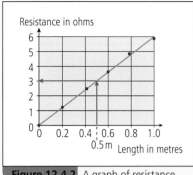

Figure 12.4.2 A graph of resistance against length for a uniform wire

Supplement

KEY POINTS

1 The resistance of a uniform wire increases if its length is increased or if its diameter is reduced.

2 The resistance of a wire depends on the material it is made from and is:

i proportional to its length,

ii inversely proportional to its area of cross-section.

SUMMARY QUESTIONS

1 Copy and complete the following sentences using words from the list below.

greater   the same as   smaller

a A wire W in a circuit has a fixed pd across it. The length of wire W in a circuit is reduced. As a result of this change:

i the resistance of wire W in the circuit will become _____,

ii the current in wire W will become _____.

b A wire X in a different circuit is replaced by a thicker wire Y of the same length and the same material.

i The resistance of wire Y is _____ than that of wire X.

ii The current in wire Y will be _____ than the current through wire X.

2 a Describe how you would make measurements on wires of different known diameters and of the same material to show that the narrower a wire is, the greater its resistance.

b A wire of resistance 5.0 Ω and length 1.0 m in a circuit is replaced with a wire of the same material and of length 0.5 m which has half the diameter of the first wire.

i A student suggests the second wire has a resistance of 2.5 Ω. Explain why this suggestion is incorrect.

ii Calculate the resistance of the second wire. Explain the steps in your calculation.

Supplement

Supplement

Tests show that the resistance of a uniform conductor is:

• proportional to its length. For example, if the length is doubled, the resistance is also doubled. The graph in Figure 12.4.2 shows the resistance is proportional to the length because the line is straight and passes through the origin.

• inversely proportional to the area of cross-section of the wire. For example, if the wire is replaced by wire of the same length and of the same material but with twice the cross-sectional area, the resistance is halved.

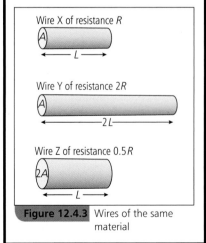

Figure 12.4.3 Wires of the same material

Supplement

LEARNING OUTCOMES

- Recall how electrical power depends on current and pd
- Calculate electrical power from current and pd
- Recall how electrical energy supplied to a device depends on its electrical power and the time in use
- Calculate electrical energy from current, pd and time

STUDY TIP

Power is a very important idea, often not understood by students. Remember that 1 watt = 1 joule per second.

DID YOU KNOW?

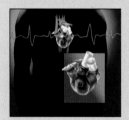

An artificial heart

The power of a human heart varies with activity and is about 1 watt on average. In 1 year, a human heart is supplied with about 30 million joules of energy. A surgeon fitting an artificial heart in a patient needs to make sure the battery will last a long time. Even so, the battery must be replaced every few years.

Energy and power

When you use an appliance, it transforms energy from one form into other forms of energy. The **power** of an appliance, in watts, is the rate at which it transforms energy, in joules per second.

$$\text{power (in watts)} = \frac{\text{energy transformed (in joules)}}{\text{time taken (in seconds)}}$$

WORKED EXAMPLE

A light bulb transforms 30 000 J of electrical energy when it is on for 300 s. Calculate its power.

Solution

$$\text{Power} = \frac{\text{energy transformed}}{\text{time taken}} = \frac{30\,000\,\text{J}}{300\,\text{s}} = 100\,\text{W}$$

Calculating power

Millions of millions of electrons pass through the circuit of an artificial heart every second. Each electron transfers a small amount of energy to it from the power supply. So the energy transferred to it each second is large enough to enable the device to work.

For any electrical appliance:

- the current in it is a measure of the number of electrons passing through it each second (i.e. the charge flow per second)
- the potential difference across it is a measure of how much energy each electron passing through it transfers to it (i.e. the electrical energy transferred per unit charge)
- the power supplied to it is the energy transferred to it each second.

Therefore,

$$\begin{array}{c}\text{the energy transfer to}\\\text{the device each second}\end{array} = \begin{array}{c}\text{the charge flow}\\\text{per second}\end{array} \times \begin{array}{c}\text{the energy transfer}\\\text{per unit charge}\end{array}$$

In other words,

$$\begin{array}{c}\textbf{power supplied}\\\text{(in watts)}\end{array} = \begin{array}{c}\textbf{current}\\\text{(in amperes)}\end{array} \times \begin{array}{c}\textbf{potential difference}\\\text{(in volts)}\end{array}$$

For example, the power supplied to:

- a 4 A, 12 V electric motor is 48 W (= 4 A × 12 V),
- a 0.1 A, 3 V torch lamp is 0.3 W (= 0.1 A × 3.0 V).

Electrical energy and potential difference

When a resistor is in a circuit, electrons are forced through the resistor by the power supply. Each electron repeatedly collides with the vibrating atoms of the resistor, transferring energy to them. The atoms of the resistor therefore gain kinetic energy and vibrate even more. The resistor becomes hotter.

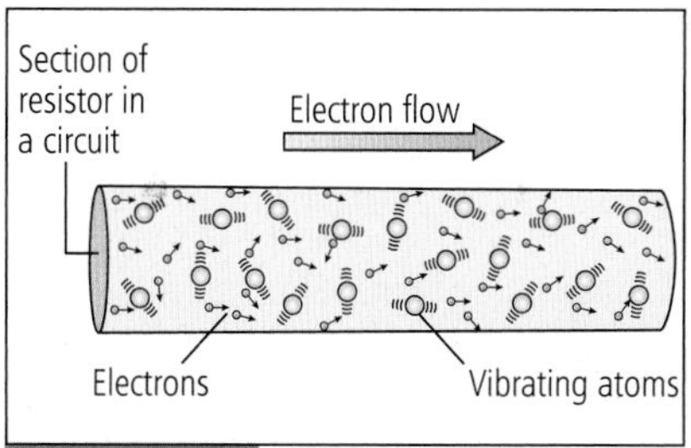

When charge flows through a resistor, electrical energy is transformed into heat energy.

The electrical energy supplied to a resistor (or any other electrical component) in a given time can be calculated if we know the pd across the resistor and the current in it. This is because:

• energy = power × time

• electrical power = current × potential difference

Therefore, we can calculate the electrical energy supplied using the equation:

electrical energy supplied	=	**current**	×	**potential difference**	×	**time**
(in joules)		(in amperes)		(in volts)		(in seconds)

$$E = IVt \qquad \text{where} \quad I = \text{current}$$
$$V = \text{potential difference}$$
$$t = \text{time taken}$$

Note

For a resistor of resistance R, the pd across it, $V = IR$ when the current in it is I.

In time t, the electrical energy supplied, $E = IVt = I^2Rt$

Therefore, the heat energy developed in the resistor = I^2Rt

Figure 12.5.1 Electrons moving through a resistor

KEY POINTS

1 Electrical power supplied (in watts)
= current (in amperes) × potential difference (in volts)

2 Energy transferred
= power × time (in seconds)
= current × pd × time

SUMMARY QUESTIONS

1 Copy and complete the sentences **a** to **c** using words from the list below.

current   energy   potential difference   power

a When an electrical appliance is on, _____ is supplied to it as a result of _____ passing through it.

b When an electrical appliance is on, a _____ is applied to it which causes _____ to pass through it.

c Charge flowing through a resistor transfers _____ to the resistor.

2 a Calculate the power supplied to each of the following devices in normal use:

   i a 12 V, 3 A light bulb

   ii a 230 V, 2 A heater

  b Calculate the energy transfer:

   i for a charge flow of 20 C when the potential difference is 6.0 V

   ii in 20 s for a current of 3 A in a resistor when the potential difference is 5 V.

1 Draw a diagram using standard circuit symbols of a circuit designed to show the relationship between the potential difference across a lamp and the current in the lamp at a variety of different potential differences.

2 (a) Explain the meaning of the term electromotive force (emf).

(b) Explain how you would measure the emf of a cell.

3 Write down the equation linking potential difference, current and resistance. State the units of each of the three quantities.

4 Use the readings in the table to plot a graph of V / V (y-axis) against I / A (x-axis).

V / V	I / A
0	0
1.8	0.33
4.0	0.68
6.2	1.00
7.9	1.33
10.0	1.70
12.1	2.00

Use the graph to calculate the resistance of the component in which the current is being measured.

5 (a) State two factors that affect the resistance of a metal wire.

(b) State whether an increase in the factor increases or decreases the resistance of the wire.

6 (a) Calculate the power of a 240 V lamp that carries a current of 1.6 A.

(b) Calculate the amount of energy supplied if the lamp is left switched on for 10 min.

7 (a) Write down the equation that links the emf of a battery with the electrical energy it produces and the amount of charge that passes through it.

(b) (i) A 12 V battery passes 48 C of charge. Calculate how much energy it produces.

(ii) Which other quantity would you need to have measured in order to calculate the power?

1 The unit of emf is the

A amp   B ohm   C volt   D watt

(Paper 1/2)

2 The current in a resistor is 0.4 A. The potential difference across the resistor is 1.8 V. The resistance, in ohms, of the resistor is

A 0.22   B 1.4   C 2.2   D 4.5

(Paper 1/2)

3 A 60 W lamp is left switched on for 10 min. The amount of energy, in J, that is used is

A 0.1

B 6.0

C 600

D 36 000

(Paper 2)

4 Consider the following properties of a wire: 1, length; 2, diameter; 3, the metal it is made from; 4, temperature.

Which of these affect the resistance of the wire?

A 1 and 2 only

B 1, 3 and 4 only

C 1, 2, 3, and 4

D 3 only

(Paper 2)

5 Which of the following graphs correctly shows the relationship between current and potential difference for a metallic conductor at constant temperature?

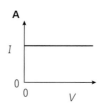

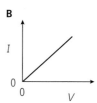

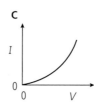

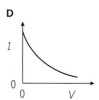

(Paper 1/2)

6 A student is carrying out an experiment to determine the resistance of a resistor. Part of the circuit diagram is shown below.

(a) (i) Copy and complete the diagram, using standard circuit symbols, to include a switch, ammeter, voltmeter and variable resistor.

(ii) What is the purpose of the variable resistor in the circuit? *[5]*

(b) The readings in the experiment are shown in the table.

V / V	0	2.1	4.0	5.9	8.1	10.0
I / A	0	0.19	0.37	0.55	0.75	0.93

(i) Plot the graph of *V* / V (*y*-axis) against *I* / A (*x*-axis). Draw the best-fit straight line.

(ii) State what the graph line shows about the resistance of the resistor.

[5]

(Paper 3)

7 (a) (i) The current in a 240 V lamp at normal brightness is 0.25 A. Calculate the power of the lamp.

(ii) State the amount of energy used by the lamp in 1 s.

(iii) Some of the electrical energy is transformed into light. To which form is most of the energy transformed? *[5]*

(b) A variable resistor could be used in a circuit to make the lamp dimmer.

(i) Draw the circuit symbol for a variable resistor.

(ii) When the resistance of the variable resistor is increased the lamp dims. Explain why the lamp dims.

(iii) Name the instrument you would use to measure the current in the lamp. *[3]*

(Paper 3)

8 If the temperature of a wire increases, its resistance increases.

(a) Copy and complete the table below with *three* other properties that affect the resistance of a wire and how the resistance is affected. The example of increase in temperature is done for you.

Property	Effect of increase in property on resistance
Temperature	Increase

[6]

(b) The graph shows the results obtained by a student who is determining the resistance of a fixed length of wire.

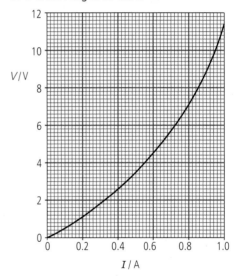

(i) Use the graph to find the current when the potential difference is 5.0 V. Show how you obtained the necessary information.

(ii) Calculate the resistance of the wire when the potential difference is 5.0 V.

(iii) The student expected the graph to be a straight line which would have shown that the resistance of the wire was constant. However the graph is a curve showing that the resistance is increasing. Suggest a reason for the resistance not being constant in this experiment. *[6]*

(Paper 4)

13 Electric circuits

13.1 Circuit components

Supplement

LEARNING OUTCOMES

- Recall commonly-used circuit symbols
- Draw and interpret circuit diagrams
- Draw and interpret circuit diagrams containing diodes

PRACTICAL

Circuit tests

1 Connect a variable resistor in series with the torch lamp and a battery, as shown in Figure 13.1.1. Adjusting the slider of the variable resistor alters the amount of current flowing through the bulb and therefore affects its brightness.

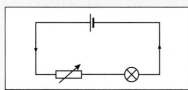

Figure 13.1.1 Using a variable resistor

2 Move the slider one way so the torch lamp becomes dim. This happens because more of the variable resistor is included in the circuit when the slider is moved this way. So the current decreases.

Supplement

3 Replace the variable resistor with a diode in its 'forward' direction so the torch lamp is on. Reverse the diode in the circuit and you should find the lamp is off. Sketch and label the circuit diagram with the diode in its forward direction.

Components and their symbols

A circuit diagram is a very helpful way of showing how the components in a circuit are connected together. Each component has its own symbol. Figure 13.1.2 shows the symbols for some of the components you will meet in this course. The function of each component is also described in Figure 13.1.2. You need to recognise these symbols and remember what each component is used for.

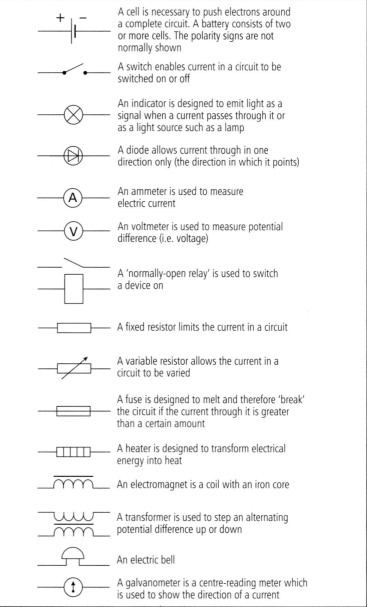

Figure 13.1.2 Components and symbols

Fuses

A fuse is designed to protect an appliance. It must be connected in series with the appliance. A fuse contains a thin wire that melts if too much current passes through it. If a fault occurs in an appliance and too much current passes through it, the fuse wire melts and the current through both it and the appliance is cut off.

The fuse for a given appliance must be chosen so it melts only when the current is greater than the 'normal' current that passes through the appliance.

Supplement

If the pd and the power of the appliance are known, the normal current can be calculated using the equation 'power = current × pd'.

For example, the normal current through an 800 W, 230 V microwave oven is 3.5 A (= 800 W/230 V). If a 3 A fuse, a 5 A fuse and a 13 A fuse were available for this appliance, the 5 A fuse should be used as the 3 A fuse would melt in normal operation and the 13 A fuse would allow too much current through before it melted.

Diodes

A diode allows current through in one direction only, referred to as its 'forward' direction. Its resistance in the forward direction is very low. Its resistance in the reverse direction is very high.

STUDY TIP

Make sure that you learn the symbols and do not confuse them with each other. A common mistake is drawing a fuse instead of a resistor.

DID YOU KNOW?

A portable radio would be damaged if the batteries were put in the wrong way around unless a diode is in series with the battery. The diode allows current through only when it is connected as shown in Figure 13.1.3. If the battery is reversed in the circuit, the diode stops electrons passing around the circuit.

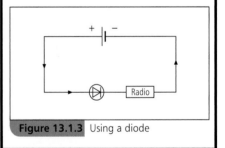

Figure 13.1.3 Using a diode

SUMMARY QUESTIONS

1 Name the numbered components in the circuit diagram in Figure 13.1.4 and state the function of each one.

2 Figure 13.1.5 shows the construction of a torch.
   a Draw the circuit diagram for Figure 13.1.5.
   b i What is the purpose of the spring in Figure 13.1.5?
     ii Why is the case of the torch made of plastic rather than metal?

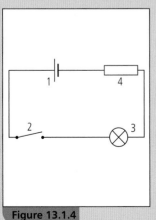

Figure 13.1.4

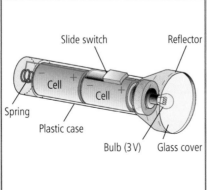

Slide switch · Reflector · Cell · Cell · Spring · Plastic case · Bulb (3 V) · Glass cover

Figure 13.1.5

KEY POINTS

1 Every component has its own agreed symbol.

2 A circuit diagram shows how components are connected together.

3 A diode allows current through in its 'forward' direction only.

Series circuits

Supplement

LEARNING OUTCOMES

- Recognise that the same current passes through each component in series
- Calculate the combined resistance of resistors in series
- Recall and use the pd rule for components in series

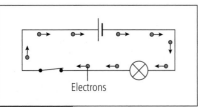

Figure 13.2.1 A torch lamp circuit

STUDY TIP

Remember the current is the same all the way around a series circuit.

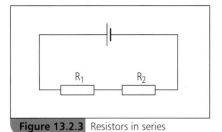

Figure 13.2.3 Resistors in series

Circuit rules

1 The same current passes through components in series with each other.

In the torch circuit in Figure 13.2.1, the lamp, the cell and the switch are connected in series with each other. Each electron moving around the circuit passes through every component in the circuit. The same number of electrons passes through each component every second, so the same current is in each component. This is true for any series circuit.

PRACTICAL

Measuring the current in a series circuit

1 Figure 13.2.2 shows a lamp, a cell, a variable resistor and two ammeters A_1 and A_2 in series with each other.

2 Set up this circuit and use the variable resistor to change the current in the circuit.

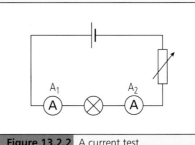

Figure 13.2.2 A current test

You should find that the two ammeters always show the same current as each other. The current is the same in every part of the circuit because the components are in series.

2 The total resistance of two or more resistors in series is equal to the sum of their separate resistances.

In Topic 12.4, we saw that the resistance of a uniform wire depends on its length. We can use the graph in Figure 12.4.2 to find the resistance of any length of wire up to 1.0 m. For example:

- a 0.4 m length of the wire has a resistance of 2.4 Ω,
- a 0.6 m length of the wire has a resistance of 3.6 Ω.

Suppose the two lengths are connected in series so together they are equivalent to a 1.0 m length of the wire. The graph gives a resistance of 6.0 Ω for a 1.0 m length. This is equal to the sum of the separate resistances of each length (i.e. 2.4 Ω for the 0.4 m length + 3.6 Ω for the 0.6 m length).

In general, for two resistors of resistances R_1 and R_2 in series:

their combined resistance $= R_1 + R_2$

3 The total potential difference across a voltage supply in a series circuit is shared between the components.

In Figure 13.2.4, each electron is pushed through the lamps by the cell. The potential difference (or voltage) of the cell is a measure of the energy transferred from the cell by each electron that passes through it. Since each electron in the circuit in Figure 13.2.4 goes through both lamps, the potential difference of the cell is shared between the lamps. In other words, in any series circuit, the total potential difference across the voltage supply is equal to the sum of potential differences across the components.

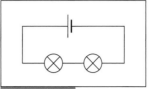

Figure 13.2.4 Lamps in series

KEY POINTS

1 For components in series:

the current is the same in each component,

the resistances add up to give the total resistance.

2 The potential differences across each component add up to give the total potential difference.

SUMMARY QUESTIONS

1 Complete the following sentences using words from the list.

greater than   less than   the same as

For the circuit in Figure 13.2.6:

a the current in the battery is _____ the current in resistor P.

b the potential difference across resistor Q is _____ the potential difference across the battery.

c The pd across resistor Q is _____ the pd across resistor P.

Two 1.5 V cells

$2\,\Omega$ $10\,\Omega$
P   Q

Figure 13.2.6

2 For the circuit in question 1, the battery has a pd of 3.0 V.

a Calculate the total resistance of the two resistors.

b Show that current in the battery is 0.25 A.

c Calculate the potential difference across each resistor.

PRACTICAL

Investigating potential differences in a series circuit

Set up the circuit shown in Figure 13.2.5 to test the potential difference rule for a series circuit. The circuit consists of a filament lamp in series with a variable resistor and a cell. Use the variable resistor to see how the voltmeter readings change when the current is changed. Make your own measurements and see how they compare with the ones in the table.

filament lamp	voltmeter V_1 / volts	voltmeter V_2 / volts
normal	1.5	0.0
dim	0.9	0.6
very dim	0.5	1.0

Your measurements should show that the voltmeter readings always add up to 1.5 V. This is the potential difference across the cell.

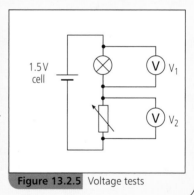

1.5 V cell

Figure 13.2.5 Voltage tests

STUDY TIP

Are your own readings correct 'within the limits of experimental accuracy'?

Supplement

LEARNING OUTCOMES

- Explain what a parallel circuit is
- Recognise that the current in each branch of a parallel circuit is less than the current from the power supply
- Recall that the current from the power supply is the sum of the currents in the branches of a parallel circuit

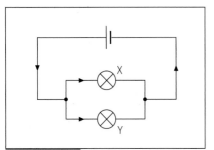

Figure 13.3.1 Torch lamps in parallel

STUDY TIP

Parallel circuits always have a junction where the current must split and take more than one route. But all the routes end up back at the power source.

Lamps in parallel

Can you design a circuit with two torch lamps, X and Y, and a cell so that one lamp stays on if the other one is disconnected? Figure 13.3.1 shows such a circuit. The two torch lamps are said to be in parallel because the current through X does not pass through Y and the current through Y does not pass through X. If X is disconnected, Y stays lit. If Y is disconnected instead, X stays lit. The circuit in Figure 13.3.1 has two parallel branches, one with lamp X in and the other with lamp Y in.

Investigating parallel circuits

Figure 13.3.2 shows how you can investigate the current in two lamps in parallel with each other. Ammeters in series with the lamps and the cell are used to measure the current in the lamp. Set up your own circuit and see how the measurements compare with the ones in the table for different settings of the variable resistor.

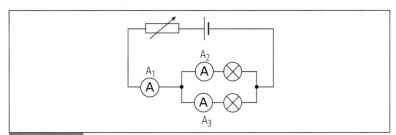

Figure 13.3.2 At a junction

ammeter A_1 / A	ammeter A_2 / A	ammeter A_3 / A
0.50	0.30	0.20
0.30	0.20	0.10
0.18	0.12	0.06

You should find that the reading of ammeter A_1 is always greater than the reading of either of the other two ammeters. This means that the current in the cell is always larger than the current in either of the two lamps. In general, for a parallel circuit:

the current from the power supply is larger than the current in each branch.

Your results from the circuit in Figure 13.3.2 should show, as in the table, that the reading of ammeter A_1 is equal to the sum of the readings of ammeters A_2 and A_3. In general, for any parallel circuit:

The current from the power supply is the sum of the currents in the separate branches of the circuit.

Supplement

A household lighting circuit

In a building, each lamp is switched on or off by a separate switch. The lamps can be switched on or off independently. When the lighting circuit is installed, the lamps and switches must be correctly connected to the lighting circuit. Otherwise, the lamps might switch on and off together or the fuses in the circuit might blow.

Figure 13.3.3 shows the circuit diagram for a household lighting circuit supplied with **mains electricity** via two wires referred to as the **live wire** and the **neutral wire**.

Each lamp is in series with a switch that switches the lamp on or off.

- Each lamp and its switch is a parallel branch of the lighting circuit connected between the live wire and the neutral wire.

- A fuse in series with the live wire cuts the mains supply off if too much current passes along the live wire. For example, if a fault occurred in one of the lamps and the current through it increased considerably, the fuse would melt and cut the lamps off from the circuit.

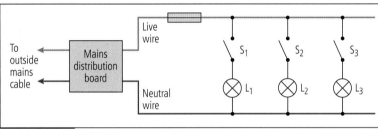

Figure 13.3.3 A household lighting circuit

STUDY TIP

Without parallel circuits in a building, the only choice of lighting would be every lamp on or every lamp off!

KEY POINTS

1 The current in each branch of a parallel circuit is less than the current from the power supply.

2 The current from the power supply is the sum of the currents in the branches of a parallel circuit.

Supplement

SUMMARY QUESTIONS

1 Figure 13.3.4 shows two circuits with identical lamps.
   a Which circuit allows each torch lamp to be switched on and off without affecting the other torch lamp?
   b In circuit 1, how could both lamps be switched on?
   c Both switches are closed in each circuit. State and explain whether the current in the cell in circuit 1 is greater than, less than or the same as the current in the other cell.

2 The circuit diagram in Figure 13.3.5 shows three resistors $R_1 = 2\,\Omega$, $R_2 = 3\,\Omega$ and $R_3 = 6\,\Omega$ connected to each other in parallel and to a 6 V battery.

   Calculate:
   a the current in each resistor,
   b the current in the battery.

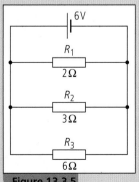

Figure 13.3.5

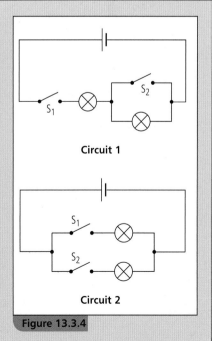

Circuit 1

Circuit 2

Figure 13.3.4

Supplement

Supplement

LEARNING OUTCOMES

- Recognise that components in parallel have the same pd
- Relate the current through a parallel component to its resistance
- Calculate the combined resistance of two resistances in parallel

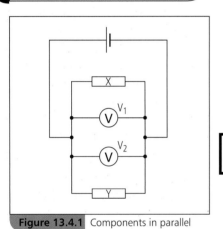

Figure 13.4.1 Components in parallel

STUDY TIP

Many students do not realise the pd across parallel components is the same and therefore misunderstand much basic electricity!

Potential differences in a parallel circuit

Figure 13.4.1 shows two resistors X and Y in parallel with each other. A voltmeter is connected across each resistor. The voltmeter across resistor X shows the same reading as the voltmeter across resistor Y. This is because each electron from the cell either passes through X or through Y. So it delivers the same amount of energy from the cell whichever resistor it goes through. In other words, for components in parallel:

the potential difference across each component is the same.

Cells in series

What happens if we use two or more cells in series in a circuit (Figure 13.4.2). Provided we connect the cells so they act in the 'same direction', each electron gets a push from each cell. So an electron would get the same push and therefore the same energy from a battery of three 1.5 V cells in series as it would from a single 4.5 V cell.

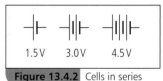

1.5 V	3.0 V	4.5 V

Figure 13.4.2 Cells in series

In other words:

the total emf of cells in series acting in the same direction is the sum of the individual emfs of the cells.

Supplement

Note: Provided no energy is wasted as heat inside a cell (i.e. the internal resistance of a cell is zero when current passes through it), the pd across the terminals of a cell is equal to its emf. In examination questions for this specification, assume cells, batteries and other power supplies have no internal resistance unless told otherwise in a question.

PRACTICAL

Measuring potential differences in a parallel circuit

1. Set up and test the circuit in Figure 13.4.1.

2. Repeat the test with two cells in series, then with three cells in series, all acting in the same direction.

3. Record your readings in a table.

	V_1 / V	V_2 / V
one cell	1.5	1.5
two cells		
three cells		

The table should show that the voltmeter readings:

- depend on the number of cells in the circuit,
- are the same in each of the three tests.

Rules for resistors in parallel

Resistors in parallel have the same potential difference. For two or more resistors in parallel:

• The current in any of the resistors depends on the resistance of the resistor. The bigger the resistance of the resistor, the smaller the current in it. The resistor which has the largest resistance passes the smallest current.

• The total current in the combination is larger than the current in any of the resistors. Therefore, the combined resistance of the resistors is less than the smallest individual resistance.

For example, Figure 13.4.3 shows a $3\,\Omega$ resistor and a $6\,\Omega$ resistor connected in parallel with a $12\,V$ battery.

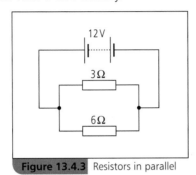

Figure 13.4.3 Resistors in parallel

• The current in the $3\,\Omega$ resistor is larger than the current in the $6\,\Omega$ resistor.

• The current in the battery is larger than the current in either resistor.

SUMMARY QUESTIONS

1 Copy and complete the sentences **a** to **c** using words from the list below.

current   potential difference   circuit

a Components in parallel with each other have the same _____.

b For components in parallel, the _____ is greatest for the component with the least resistance.

c For two unequal resistors in parallel connected to a cell, the _____ is not the same for each resistor.

2 A circuit consists of a $4.0\,\Omega$ resistor in parallel with a $12.0\,\Omega$ resistor connected to a $12.0\,V$ battery.

a i Draw the circuit diagram.

ii State and explain which resistor current has the greater current passing through it.

b Calculate:

i the effective resistance of the two resistors,

ii the current through each resistor,

iii the current through the battery.

Supplement

Supplement

In Figure 13.4.3:

the current in the

$3\,\Omega$ resistor $= \dfrac{12\,V}{3\,\Omega} = 4\,A$

the current through the

$6\,\Omega$ resistor $= \dfrac{12\,V}{3\,\Omega} = 2\,A$

Therefore, the current in the cell $= 4\,A + 2\,A = 6\,A$

For a current of $6\,A$ to pass through a single resistor connected to a $12\,V$ battery, the resistance would need to be $2\,\Omega$ ($=$ pd/current $= 12\,V/6\,A$). Therefore, the parallel combination has a combined resistance of $2\,\Omega$.

Using the same approach as above, it can be shown that the combined resistance R of two or more resistors of resistances R_1 and R_2 in parallel is given by:

$$\frac{1}{R} = \frac{1}{R_1} + \frac{1}{R_2}$$

Prove for yourself that the combined resistance of a $3\,\Omega$ resistor in parallel with a $6\,\Omega$ resistor is $2\,\Omega$.

KEY POINTS

1 For components in parallel:

The potential difference is the same in each component.

The total current is the sum of the currents through each component.

The bigger the resistance of a component, the smaller the current.

2 The combined resistance R of two resistances R_1 and R_2 in parallel is given by:

$$\frac{1}{R} = \frac{1}{R_1} + \frac{1}{R_2}$$

Sensor circuits

LEARNING OUTCOMES

- Describe a potential divider
- Describe the action of a variable potential divider
- Explain how a thermistor works in a temperature sensor circuit
- Explain how an LDR thermistor works in a light sensor circuit

STUDY TIP

The potential divider seems difficult at first. But persevere – it's not as difficult as it appears!

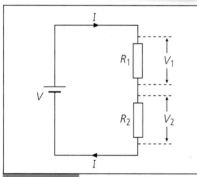

Figure 13.5.1 The potential divider

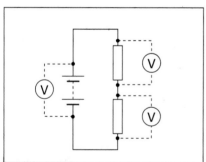

Figure 13.5.2 Investigating the potential divider

The potential divider

Sensor circuits are designed to respond to an external change such as a change of temperature or light intensity. Many sensor circuits consist of a potential divider which includes a resistor that is sensitive to a change in the surroundings.

A potential divider consists of two or more resistors in series. A fixed pd is connected across the combination, as shown in Figure 13.5.1. The pd across the combination is shared or 'divided' between the resistors.

Figure 13.5.1 shows two resistors of resistances R_1 and R_2 connected to a source of fixed pd V. Because the resistors are in series, the same current passes through the two resistors. The share across each resistor of the fixed pd depends on the resistance of the resistor in relation to the total resistance.

For example, for two resistors $R_1 = 10\,\Omega$ and $R_2 = 40\,\Omega$ in series and a current $I = 0.1\,A$:

- the pd, V_1, across the $10\,\Omega$ resistor $= IR_1 = 0.1\,A \times 10\,\Omega = 1.0\,V$
- the pd, V_2, across the $40\,\Omega$ resistor $= IR_2 = 0.1\,A \times 40\,\Omega = 4.0\,V$.

Because the resistors are in series with each other, the pd across the combination $= 1.0\,V + 4.0\,V = 5.0\,V$

Therefore, the fixed pd of 5.0 V across the combination is shared between the two resistors as stated above.

PRACTICAL

Investigating the potential divider

1 Connect a battery to two resistors of known resistance in series with each other, as shown in Figure 13.5.2. Use a voltmeter to measure the battery pd and the pd across each resistor. You should find that the sum of the pds across the resistors is equal to the battery pd.

2 Replace one of the resistors with a variable resistor and measure the pd across the fixed resistor for several different settings of the variable resistor. You should find that the pd across the fixed resistor varies between zero and the battery pd according to the setting of the variable resistor. For each setting of the variable resistor, the pd across the variable resistor is equal to the difference between the battery pd and the pd across the fixed resistor.

A **potentiometer** is a variable potential divider consisting of a track of a suitable resistive material with a fixed contact at each end and a sliding contact on the track, as shown in Figure 13.5.3. The track is straight in a linear potentiometer and circular in a rotary potentiometer.

In Figure 13.5.3, the voltmeter reading:

• increases when the sliding contact is moved up

• decreases when the sliding contact is moved down.

The pd between the sliding contact and either fixed contact depends on the position of the sliding contact.

Input transducers

We use thermistors and light-dependent resistors (LDRs) in **sensor circuits**. A change of the output pd of a sensor circuit can be used to activate an alarm or to control a device. A sensor circuit is sometimes described as an **input transducer**.

A **thermistor** is a temperature-dependent resistor. Its resistance decreases the hotter it becomes. The symbol for a thermistor is shown in Figure 13.5.4.

Figure 13.5.4 shows a sensor circuit in which a thermistor T and a resistor R form a potential divider connected to a battery. When the temperature of the thermistor increases, its resistance decreases so its share of the battery pd decreases. As a result, the share of the battery pd across resistor R increases so the output pd increases.

An **LDR** has a resistance that decreases if the brightness of the incident light increases. The symbol for an LDR is shown in Figure 13.5.5.

When the brightness of the incident light is increased, its resistance decreases so the output pd increases as explained above.

In both of the sensor circuits, the resistor may be replaced with a variable resistor which is adjusted to give a certain output pd for a certain temperature or light brightness. The output pd would then change if the temperature or the light brightness changes.

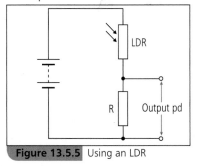

Figure 13.5.5 Using an LDR

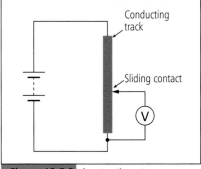

Figure 13.5.3 A potentiometer

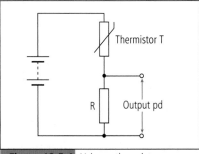

Figure 13.5.4 Using a thermistor

KEY POINTS

1 A potential divider consists of two resistors in series connected to a fixed pd.

2 A thermistor is a resistor which has a resistance that decreases when its temperature is increased.

3 An LDR is a resistor which has a resistance that decreases when the incident light is made brighter.

SUMMARY QUESTIONS

1 A potential divider consisting of a 20 Ω resistor in series with a 30 Ω resistor is connected to a 5.0 V battery.

a Sketch the circuit diagram and explain why the pd across the 30 Ω resistor is greater than the pd across the 20 Ω resistor.

b Calculate the current through the battery and the pd across the 30 Ω resistor.

2 a In Figure 13.5.5, explain why the output pd decreases when the LDR is covered completely so it is in darkness.

b The thermistor and the resistor in Figure 13.5.4 are swapped over in the circuit so the output pd is across the thermistor.

i Draw the new circuit diagram.

ii State and explain how the output pd in the new circuit changes when the temperature of the thermistor is reduced.

Supplement

LEARNING OUTCOMES

- Describe the action of a relay
- Recognise the use of a relay and an LDR or a thermistor in a switching circuit

STUDY TIP

You need to know about magnets, electricity and electromagnetism to understand the relay – good for revision!

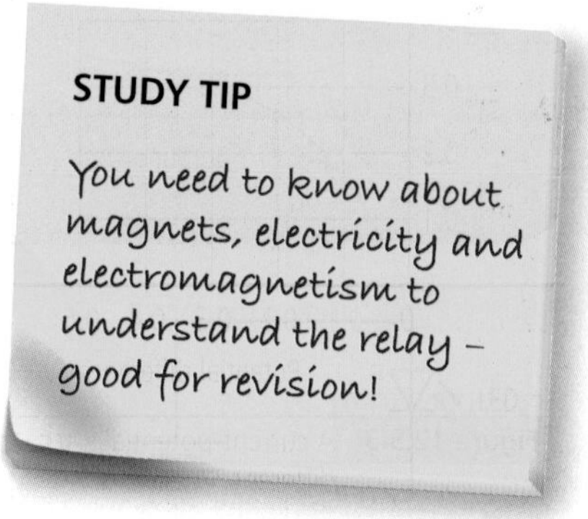

The relay

Relays are used in control circuits to switch machines on or off. Figure 13.6.1 shows the construction of a relay.

- When a current passes through the coil of the electromagnet, the iron armature is attracted on to the electromagnet.
- The armature turns about a pivot and closes the switch gap.

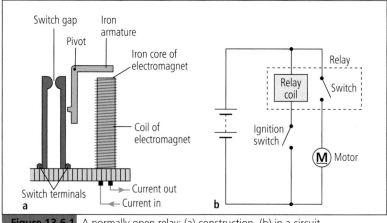

Figure 13.6.1 A normally open relay: (a) construction, (b) in a circuit

In this way a small current can be used to switch on a much greater current. For example, when the ignition switch of a car is turned on, a small current passes through the electromagnet coil so the relay switch closes. This allows a much greater current to pass through the starter motor.

Supplement

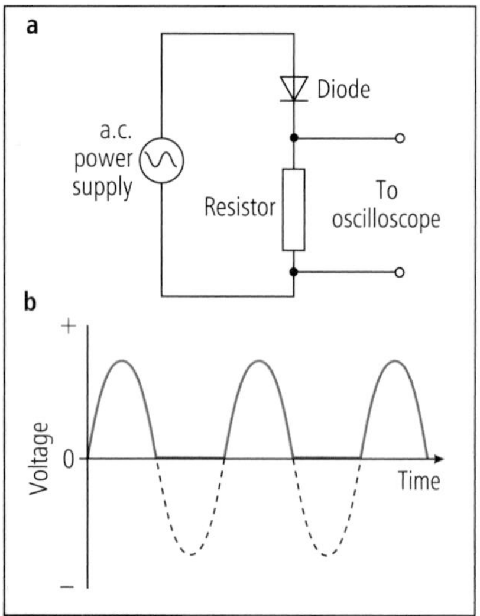

Converting alternating to direct current

- **Rectifier circuits** are circuits that convert alternating current to direct current. Such a circuit includes one or more diodes. Figure 13.6.2a shows a half-wave rectifier circuit.
- The diode allows current to pass around the circuit only when the pd across it is in its 'forward' direction. The diode is said to 'rectify' the alternating current. The 'half-wave' variation of the pd across the resistor may be displayed on an oscilloscope as shown in Figure 13.6.2b.

Figure 13.6.2 (a) Half-wave rectification
(b) a half-wave

A temperature-operated fan

Figure 13.6.3 shows a circuit which is used to switch an electric fan on if the temperature of a thermistor increases above a certain level.

When the thermistor is cold, its resistance is high so the pd across the relay coil is too small to switch the relay on.

1 Suppose the variable resistor is then adjusted so the relay is only just switched off.

2 If the thermistor temperature then increases, its resistance decreases and the pd across the relay coil therefore increases, thus switching on the relay.

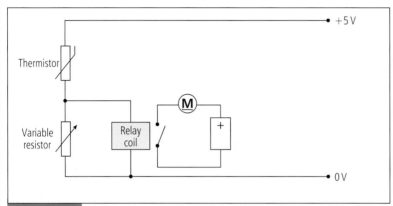

Figure 13.6.3 A temperature-operated fan

SUMMARY QUESTIONS

1 Copy and complete the sentences **a** and **b** using words from the list below.

   armature

   coil

   electromagnet

   switch

   a When a current passes through the _____ of a relay, the _____ attracts the _____.

   b The movement of the _____ closes the _____ of the relay.

2 a In Figure 13.6.3, explain the function of:

   i   the variable resistor

   ii   the relay coil

   b In Figure 13.6.3, the fan motor is replaced by a lamp.

   i   What further changes would need to be made to the circuit in order to switch the lamp on automatically at night?

   ii   Draw a circuit diagram to show the changes.

KEY POINTS

1 A relay is a switch operated by an electromagnet.

2 A potential divider containing a thermistor or an LDR can be used with a relay to switch a device on and off.

Supplement

LEARNING OUTCOMES

- Explain what a digital circuit is and what an analogue circuit is
- Describe the truth tables for different logic gates

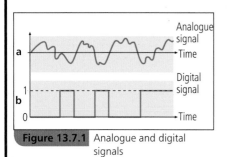

Figure 13.7.1 Analogue and digital signals

INPUTS		OUTPUT
Am I the only person in at the moment?	Will I be out for more than a few minutes?	Should I lock up?
0	0	0
0	1	0
1	0	0
1	1	1

1 = YES; 0 = NO.

Digital electronics

Which electronic devices do you use every day? Your list will probably include a mobile phone, a home computer and an iPod or an MP3 player. All these devices are **digital** devices. The potential or voltage at any point in a digital circuit is either zero (i.e. low) or at a fixed positive value (i.e. high) with reference to the negative terminal of the power supply. No other values are possible in a digital circuit. A digital signal is a sequence of pulses. The voltage level of each pulse is either high (a '1') or low (a '0') with no in-between levels.

The digital 'age' in which we now live became possible when physicists discovered, about 40 years ago, how to make switching circuits containing transistors on a single small 'chip' of silicon. Before the invention of the silicon chip, most electronic devices including phones were **analogue** devices. The voltage at any point in an analogue circuit could vary between the maximum and minimum voltage of the power supply as shown in Figure 13.7.1.

Electronic logic

Decisions are made by each of us all the time. A logical decision is one with an outcome that depends on the 'input' conditions. For example, suppose you are thinking about whether or not you should lock up your home when you go out. The table on the left shows the possible 'input' conditions and the outcome or 'output' for each input condition. Each input and output 'state' is entered as 0 for 'No' and 1 for ' Yes' in the table. Such a table is referred to as a truth table.

Gate	Symbol	Function (High voltage = 1, Low voltage = 0)	Truth Table INPUTS OUTPUT		
			A	B	
OR	A─┐ OUTPUT B─┘	OUTPUT = 1 if A OR B = 1	0 0 1 1	0 1 0 1	0 1 1 1
AND	A─┐ OUTPUT B─┘	OUTPUT = 1 if A AND B = 1	0 0 1 1	0 1 0 1	0 0 0 1
NOR	A─┐ OUTPUT B─┘	OUTPUT = 0 if A OR B = 1	0 0 1 1	0 1 0 1	1 0 0 0
NAND	A─┐ OUTPUT B─┘	OUTPUT = 0 if A AND B = 1	0 0 1 1	0 1 0 1	1 1 1 0
NOT	OUTPUT INPUT	OUTPUT = 1 if INPUT = 0 OUTPUT = 0 if INPUT = 1	0 1		1 0

Figure 13.7.2 Logic gates

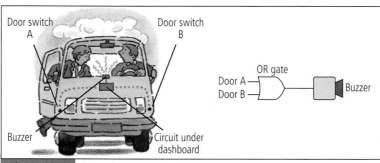

Figure 13.7.3 A warning system

A **logic gate** is a digital circuit designed to make a decision. It contains electronic switches and has one or more input terminals and an output terminal. The voltage at the output terminal (i.e. the output state) depends on the voltage at the input terminals (i.e. the input states) according to the type of gate and can be displayed using a truth table. Figure 13.7.2 shows the circuit symbol and the truth table of some common logic gates.

An example of a logic gate in use is a door alarm for the two-door car shown in Figure 13.7.3. The alarm will sound if either door is open. Figure 13.7.3 shows the logic circuit; the OR gate's truth table is in Figure 13.8.2. The logic circuit consists of a single OR gate connected to a buzzer. If door A OR door B is open, the buzzer is on.

Logic gate combinations are used for more complicated situations. For example, a 2-door car alarm that sounds only when one of the two doors is opened and the engine is running would be better than the one shown in Figure 13.7.3 for many car-users. Figure 13.7.4 shows the truth table and a suitable logic gate combination for this purpose. The truth table can be summed up by the condition

the output = 1 if input A OR input B = 1 AND input C = 1

INPUT			OUTPUT
A	**B**	**C**	
0	0	0	0
1	0	0	0
0	1	0	0
1	1	0	0
0	0	1	0
1	0	1	1
0	1	1	1
1	1	1	1

Truth table for Figure 13.7.4.

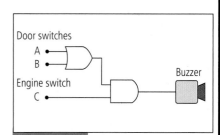

Figure 13.7.4 A logic gate combination

STUDY TIP

Truth tables are fun and easy to do.

SUMMARY QUESTIONS

1 Draw the symbol and the truth table for

  a an AND gate

  b a NOT gate

  c A NOR gate.

2 Copy and complete the following truth table for each logic gate combination in Figure 13.7.5.

INPUTS		OUTPUT
A	**B**	
0	0	
1	0	
0	1	
1	1	

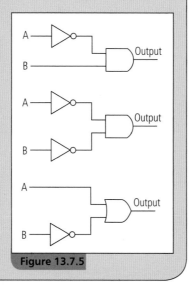

Figure 13.7.5

KEY POINTS

1 The voltage at any point in:

  • a digital circuit is either high (1) or low (0)

  • an analogue circuit can be at any value between the minimum and maximum voltage of the power supply.

2 A logic gate is a digital circuit that gives an output voltage determined by the input voltages.

LEARNING OUTCOMES

- Describe a simple digital circuit that includes logic gates
- Explain the operation of a simple digital circuit
- Design a simple digital circuit that includes several logic gates

STUDY TIP

If you concentrate on the three parts of the system, each device is easier to understand.

The car door systems described in Topic 13.7 use switches as input sensors. These sensors detect when a door is open and send a signal to the logic circuit. The circuit controls the alarm buzzer. The buzzer is an output device designed to convert an electrical signal into sound. The whole system is designed in three parts: the input sensors, the control circuit and the output device. Many electronic systems operate in this way.

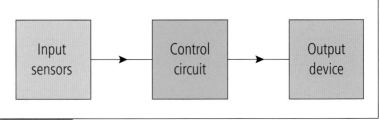

Figure 13.8.1 A control system

Input sensors

A sensor is a device that produces an electrical signal in response to a change of a physical variable such as temperature, light intensity, pressure, moisture or a magnetic field. We saw in Topic 13.5 how to use a thermistor in a potential divider to make a temperature sensor and how to use a light-dependent resistor to make a light sensor in the same way. Figure 13.8.2 shows how we can use a NOT gate to convert the output pd of such potential dividers into a digital signal.

- When the temperature of the thermistor increases, its resistance decreases so the output pd from the potential divider increases. This output pd provides the input signal to the NOT gate.

- When the thermistor temperature reaches a certain value, the output pd of the potential divider is large enough to make the output of the NOT gate switch from 1 to 0.

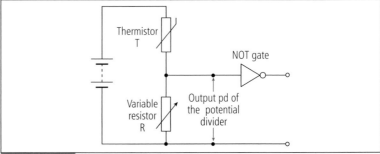

Figure 13.8.2 A digital sensor

The variable resistor can be used to change the temperature at which the output of the NOT gate switches.

We can make different types of digital sensor using other suitable components. For example:

- a light sensor would require a light-dependent resistor in place of the thermistor in Figure 13.8.2
- a pressure switch in series with a resistor could replace the thermistor in Figure 13.8.2 to make a pressure sensor.

As explained on the next page, a digital sensor can be fitted with an LED indicator to show its output state. We don't need to see the details of how a sensor works; we just need to know what physical change makes its output state change from 0 to 1 or 1 to 0. In the systems described below, we will represent a sensor as shown in Figure 13.8.3.

Control circuits

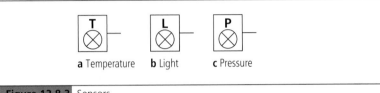

a Temperature　　**b** Light　　**c** Pressure

Figure 13.8.3　Sensors

We can use logic gates and sensors to tell us when a physical variable (e.g. temperature) is too high or too low and, if necessary, to switch a device on. The two examples below indicate the general principles.

A high-temperature indicator is shown in Figure 13.8.4. This could be used to warn if a room in a building is too hot.

The output of the OR gate is 1 when S = 1 or when T = 1. This means the indicator is on whenever the temperature is high (i.e. T = 1) or when the test switch S is closed or both.

INPUTS		OUTPUT
Test switch S open = 0 closed = 1	Temperature sensor T cold = 0 hot = 1	indicator = 0 OFF = 1 ON
0	0	0
1	0	1
0	1	1
1	1	1

Truth table for Figure 13.8.4

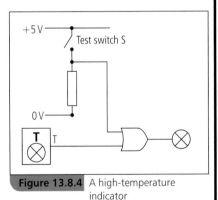

Figure 13.8.4　A high-temperature indicator

CONTINUED

STUDY TIP

Once again, if you can understand the principles of the examples given you could apply your understanding to examples you have not seen before.

A night-time rain alarm is shown in Figure 13.8.5. This could be useful if you leave some wet clothes outdoors at night to dry off in hot weather. In darkness, the light sensor L supplies a 0 to the NOT gate which therefore supplies a 1 to one of the inputs of the AND gate. If it rains, the moisture sensor M supplies a 1 to the other AND gate input. So the alarm buzzer sounds if it is dark and wet outdoors.

The truth table has an extra column for the logic state at X, the output of the NOT gate. We can say that X = NOT L. The indicator state is therefore the result of applying M and NOT L to the AND gate. In other words, the indicator is on if M = 1 AND L = 0.

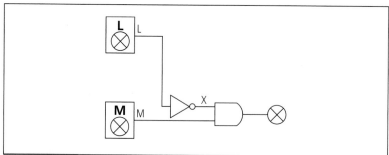

Figure 13.8.5 A night-time rain alarm

INPUTS			OUTPUT
Light sensor L dark = 0 bright = 1		Moisture sensor M dry = 0 wet = 1	indicator = 0 OFF = 1 ON
L	X		
0	1	0	0
1	0	0	0
0	1	1	1
1	0	1	0

Truth table for Figure 13.8.5

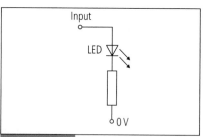

Figure 13.8.6 A logic indicator

A logic indicator

Logic indicators can be used to show the voltage state at any point in a digital circuit. For example, a logic indicator can be used to show the output state of a logic gate or of a sensor. Figure 13.8.6 shows a logic indicator consisting of a light-emitting diode (LED) and a resistor. The LED lights up when a '1' is applied to the input terminal of the indicator.

SUMMARY QUESTIONS

1 Copy and complete the truth table opposite for each logic gate combination in Figure 13.8.7.

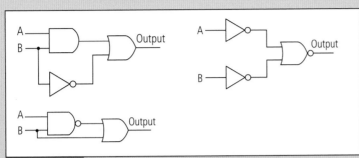

Figure 13.8.7

INPUTS		OUTPUT
A	**B**	
0	0	
1	0	
0	1	
1	1	

2 A burglar alarm system fitted in an apartment is designed so that the alarm is activated if a key operated switch is 'on' and the entrance door is open or a pressure pad behind the door is activated. Figure 13.8.8 shows the system, and part of its truth table is shown below.

  a Copy and complete the truth table.

  b Anyone entering the apartment has 20 seconds to turn the key switch off after opening the door or else the alarm is activated. Why is this delay necessary?

  c Why is the pressure pad essential?

  d Show how a two-input OR gate and a two-input AND gate may be connected together as the control circuit.

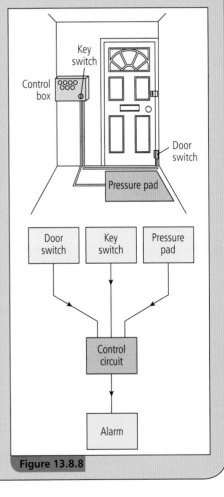

Figure 13.8.8

INPUTS			OUTPUT
Door switch	**Key switch**	**Pressure pad**	**Alarm**
OPEN = 1 CLOSED = 0	ON = 1 OFF = 0	ON = 1 OFF = 0	ON = 1 OFF = 0
0	0	0	0
1	0	0	
0	0	1	0
1	0	1	
0	1	0	0
1	1	0	
0	1	1	1
1	1	1	

KEY POINTS

1 Input sensors are used to supply a voltage signal to a control circuit.

2 A control circuit includes one or more logic gates.

3 The output signal from a control circuit is used to operate a specific device.

4 A logic indicator may be used to show the voltage state at any point in a digital circuit.

LEARNING OUTCOMES

- Explain why electrical circuits can be dangerous
- Describe common electrical faults and their causes
- Describe the features of electrical wires and fittings that make them safe to use

DID YOU KNOW?

Tap water conducts electricity. Never use an electrical appliance if you have wet or damp hands. The contact resistance between your hand and a metal terminal is much lower if your hand is wet or damp. Less resistance means more current.

Brass is used to conduct electricity in plugs

Electrical hazards

High voltage circuits are dangerous for obvious reasons. Anyone touching a bare wire or terminal at high voltage would receive an electric shock that is likely to be fatal. An electric current through the body would stop the heart. A current of no more than 0.02 A (= 20 mA) is enough to electrocute someone. The resistance of the human body is no more than about 1000 ohms. For a current of 0.02 A and a resistance of 1000 ohms, using '$V = IR$' gives a potential difference of 20 V. **Never** touch a wire or terminal at a voltage of about 20 V or more!

Low voltage circuits can be dangerous due to overheating leading to a fire. This might happen if a fault develops and causes a very large current in part of the circuit, which will then overheat. For example, suppose a wire attached to a terminal breaks off and its bare end touches a different part of the circuit. The result may be a low-resistance path for current, usually referred to as a **short-circuit**. This would allow a very large current to pass through the wire and components in series with it.

The electrical devices we use and the circuits they are connected to are all designed and manufactured to be safe.

For example:

- all wires are made of copper inside flexible hard-wearing plastic insulation. Copper is used because it is a very good electrical conductor, it does not rust and it is flexible so it doesn't snap.
- a cable used to connect a device such as an electric kettle to a socket has an outer layer of insulation surrounding the separate insulated wires inside the cable.
- plugs and sockets are made of stiff heat-resistant plastic materials shaped to hold the wires and terminals sealed firmly inside so they cannot make contact with each other.
- terminals used in plugs, sockets and other electrical fittings are made of brass because brass is a good conductor. It is more hard-wearing than copper and it does not rust or oxidise.

Common faults that develop in electrical devices and circuits have many causes including wear and tear, overloading a socket, fitting an incorrect fuse and ignoring manufacturers' instructions. Electrical faults may result in an exposed bare wire which would be lethal to touch or a short circuit if two bare wires in a cable touch each other.

Common faults include the following:

- **damaged insulation**

  The layer of insulation around a wire or a cable may wear away.

  A plug or socket may become chipped or broken, exposing a bare wire or terminal in the plug.

- **overheating of cables**

Too much current through the wires of a cable may cause overheating which could make the insulation of the wires soften or melt, exposing a bare wire in the cable. The wires inside the cable might also short-circuit causing a fire.

A coiled cable in use could overheat because it retains heat instead of losing heat to its surroundings (as it would if it was not coiled up).

- **damp or wet conditions**

Dampness in a device, a socket or a plug could cause a short-circuit inside the device or it could provide a conducting path to the outer surface. Anyone touching the device could then suffer an electric shock.

Water in a device would also cause a short-circuit and a conducting path to outside the device, as outlined above.

- **unsuitable cables and sockets**

A cable that is too long is a hazard because someone could trip over it. In addition, it may become a hazard if part of it is coiled up when it is in use.

A socket with more than one appliance connected to it may be overloaded if too much current passes through it and the cables connected to it.

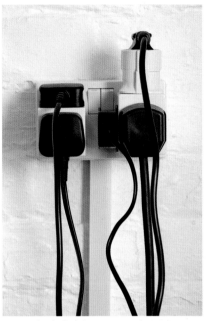

An overloaded socket

SUMMARY QUESTIONS

1 a Match the list of parts 1–4 below in a electrical plug with the list of materials A–D.

1 cable insulation	A brass
2 case	B copper
3 pin	C rubber
4 wire	D stiff plastic

  b For each part 1–4, give a reason for the choice of material.

2 a Why is a frayed or worn mains cable dangerous?

  b i Why do the wires of an electrical cable become warm if too much current passes through them?

    ii Why could a cable become dangerous if too much current passes through it?

  c Why should you never use a mains appliance if you have damp or wet hands?

STUDY TIP

Make sure you can explain why each fault is dangerous.

KEY POINTS

1 Electric circuits are dangerous because of the risk of electrocution (in high-voltage circuits) and/or overheating.

2 Common electrical hazards include damaged insulation, overheating of cables, damp conditions and overloaded sockets.

More about electrical safety

LEARNING OUTCOMES

- Describe how a fuse and a circuit breaker work
- Explain the advantage of a circuit breaker over a fuse
- Appreciate why a mains appliance with a metal case is earthed

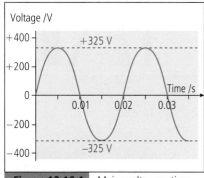

Figure 13.10.1 Mains voltage v. time

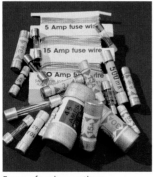

Fuses of various ratings

STUDY TIP

An alternating current repeatedly reverses its direction whereas a direct current is in one direction only. Make sure you know the difference between them.

Mains circuits

Every mains circuit has:

- a live wire which alternates from positive to negative voltage and back every cycle, and
- a neutral wire which stays at a potential or voltage close to zero because it is earthed (i.e. connected to ground) at the local sub-station. The current in a mains circuit is an **alternating current** because it reverses its direction each time the voltage reverses.

Figure 13.10.1 shows how the voltage (relative to the Earth) of the live wire varies with time.

The maximum voltage of the live wire is different in different parts of the world. In many parts of the world, it is about 325 V so the voltage of the live wire alternates between + 325 V and −325 V. In terms of electrical power, this is equivalent to a direct voltage of 230 V. So we say the 'voltage' of the mains is 230 V.

The frequency of the mains supply (i.e. the number of cycles per second) is 50 Hz in most countries except those in North America, parts of South America and a few other parts of the world. At 50 Hz, one complete cycle takes 0.02 s (= 1/50th of a second) as shown in Figure 13.10.1.

Fuses and circuit breakers

Fuses and circuit breakers are used to protect appliances and cables.

A fuse contains a thin wire that heats up and melts if too much current passes through it. The rating of a fuse is the maximum current that can pass through it without melting the fuse wire. If the rating is too large, the fuse will not blow when it should. The heating effect of the current could make the appliance catch fire. If the rating of the fuse is too low, the fuse will blow every time the appliance is switched on.

A circuit breaker is an electromagnetically-operated switch in series with an electromagnet. The switch is held closed by a spring. If the current exceeds a certain value, the electromagnet pulls the switch open (i.e. the switch 'trips') so the current is cut off. The switch stays open until it can be reset once the fault that made it trip has been put right. They work faster than fuses and can be reset quicker.

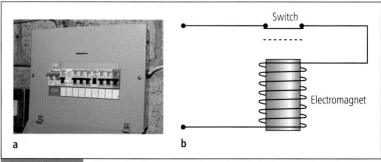

Figure 13.10.2 Circuit breakers: (a) circuit breakers in use, (b) inside a standard circuit breaker

Earthing

Mains appliances with metal frames and panels are made safer by earthing the frame. This is done automatically when the appliance is plugged in to a socket if:

- a cable is used that includes a third wire, the 'earth wire', and a suitable plug is used, and
- the circuit wiring from the distribution board includes an 'earth' wire in addition to the live and the neutral wire. The earth wire is connected to the ground outside the building.

Figure 13.10.3 shows why an electric heater is made safer by earthing its frame.

In **a**, the heater works normally and its frame is earthed. The frame is safe to touch.

In **b**, the earth wire is broken. The heater element has touched the unearthed frame so the frame is live. Anyone touching it would be electrocuted. The fuse provides no protection to the user because a current of no more than 20 mA can be lethal.

Suppose in **b** that the earth wire has been repaired but the heater element still touches the frame. The current is greater than normal and goes to earth and should blow the fuse. Because the frame is earthed, anyone touching it would not be electrocuted. But the heater could still be dangerous because the current might not be enough to blow the fuse. The appliance might therefore overheat.

Note A mains appliance with a plastic case has a 'double insulation' case made of tough, stiff plastic carrying the double insulation symbol (the double square) shown in the photograph.

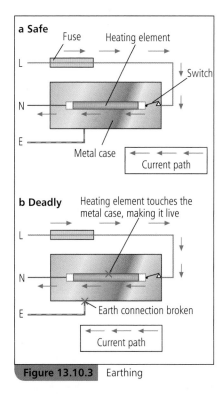

Figure 13.10.3   Earthing

The double insulation symbol used on an appliance

SUMMARY QUESTIONS

1 a What is the purpose of a fuse in a mains circuit?

  b Why is the fuse of an appliance always on the live side?

  c What advantage does a circuit breaker have compared with a fuse?

2 Figure 13.10.4 shows the circuit of an electric heater that has been wired incorrectly.

  a Does the heater work when the switch is closed?

  b When the switch is open, why is it dangerous to touch the element?

  c Redraw the circuit correctly wired.

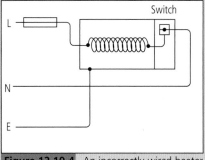

Figure 13.10.4   An incorrectly-wired heater

KEY POINTS

1 Mains electricity is an alternating current supply. A mains circuit has a live wire, which is alternately positive and negative every cycle, and a neutral wire at zero volts.

2 A fuse contains a thin wire that heats up and melts and so cuts off the current if too much current passes through it.

3 A circuit breaker is an electromagnetic switch that opens (i.e. 'trips') and cuts the current off if too much current passes through it.

1 Draw the circuit symbols for:

   (a) a buzzer

   (b) a thermistor

   (c) a diode

   (d) an LDR

   (e) a normally open relay

   (f) a variable resistor.

2 Draw a circuit diagram showing a potential divider consisting of a 50 Ω resistor and a thermistor connected to a 9.0 V battery.

3 (a) List four common faults in electrical circuits that are a safety hazard. Explain the possible danger of each one.

   (b) Describe the action of a fuse.

   (c) State the importance of an Earth wire in an electrical circuit.

   (d) Explain how an Earth wire protects a user if the metal casing of an appliance becomes live.

4 The circuit diagram shows a light-operated buzzer.

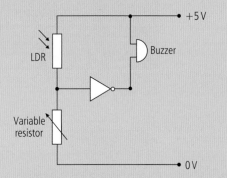

   (a) Describe the effect on the resistance of the LDR if the light shining on it changes from very dim to bright.

   (b) Explain how changing light conditions cause the buzzer to sound.

   (c) Describe the effect on the operation of the circuit of decreasing the resistance of the variable resistor.

1 The combined resistance of a 2 Ω resistor and a 8 Ω resistor in parallel is

   A 1.6 Ω

   B 5.0 Ω

   C 6.0 Ω

   D 10 Ω

(Paper 1/2)

2 The diagram shows a parallel circuit with identical resistors.

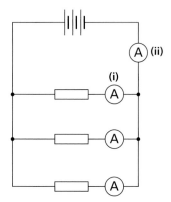

The reading on ammeter **(i)** is 0.6 A. The current reading on ammeter **(ii)** is

   A 0.2 A

   B 0.3 A

   C 0.6 A

   D 1.8 A

(Paper 1/2)

3 Three 1.5 V cells are wired together in series, acting in the same direction. The same three cells are then wired together in parallel, also acting in the same direction. The total emf across the three cells is

   A 1.5 V in series and 1.5 V in parallel

   B 1.5 V in series and 4.5 V in parallel

   C 4.5 V in series and 1.5 V in parallel

   D 4.5 V in series and 4.5 V in parallel

(Paper 2)

Supplement

4 The truth table below is for a logic gate.

Input A	Input B	Output
0	0	0
0	1	0
1	0	0
1	1	1

The type of logic gate is

A AND

B NOT

C NOR

D OR

(Paper 2)

5 The diagram shows a circuit with two resistors.

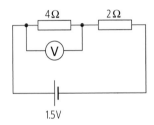

The reading on the voltmeter will be

A 0.5 V

B 1.0 V

C 1.5 V

D 6.0 V

(Paper 1/2)

6 (a) (i) Draw a circuit diagram, using standard symbols of the circuit you would use to measure the current in a series circuit. The circuit must include two lamps and an ammeter. The power source and two other components are drawn for you.

(ii) Name the other two components that are drawn for you in the diagram. *[4]*

(b) The two lamps are replaced by two resistors both with resistance of 6 Ω. Calculate the combined resistance of the two resistors. *[1]*

(Paper 3)

7 (a) Draw the circuit symbol for
(i) a thermistor and **(ii)** an LDR. *[2]*

(b) For of each of the components in part **(a)** explain the effect that causes the resistance to decrease. *[4]*

(c) Suggest one use for each of the components in part **(a)**. *[2]*

(Paper 3)

8 (a) (i) Calculate the current in a 100 W lamp connected to a 240 V supply.

(ii) Calculate the number of such lamps that could be connected in parallel to the 240 V supply safely with a 5 A fuse fitted in the circuit. *[4]*

(b) The diagram shows a household lighting circuit.

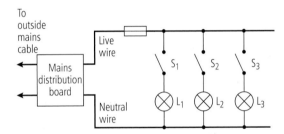

(i) Explain the importance of the fuse and describe how it works.

(ii) Describe the main advantage of the parallel circuit used. *[4]*

(Paper 4)

9 (a) Explain the terms *digital* and *analogue*. *[2]*

(b) (i) Describe the action of a diode in a circuit.

(ii) Explain how a diode can be used in a rectifier circuit. *[3]*

(c) Draw the circuit symbols for the following logic gates:

(i) NOT

(ii) AND

(iii) OR

(iv) NAND

(v) NOR *[5]*

(Paper 4)

14 Electromagnetism

14.1 Magnetic field patterns

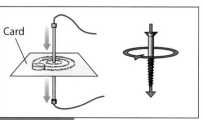

LEARNING OUTCOMES

- Describe the magnetic field pattern around a wire and a solenoid carrying a current
- Describe applications of the solenoid
- **Supplement** State how the strength and direction of the above fields vary with position

When an electric current passes along a wire, a magnetic field is set up around the wire. Figure 14.1.1 shows how the pattern of the magnetic field around a long straight wire can be seen using iron filings or a plotting compass. The lines of force due to a straight current-carrying wire are circles, centred on the wire. The field is strongest near the wire.

The direction of the field is reversed if the direction of the current is reversed. We can use the corkscrew rule to remember the direction of the magnetic field for each direction of the current. If the current is downwards, the field turns in the same direction as a downward-moving corkscrew or screw. Increasing the current increases the strength of the field everywhere in the field.

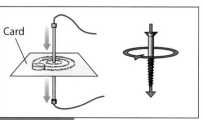

Figure 14.1.1 The magnetic field near a long straight wire

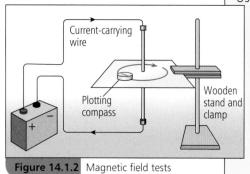

Figure 14.1.2 Magnetic field tests

PRACTICAL

Plotting the magnetic field near a current-carrying wire

1. Set up the arrangement as shown in Figure 14.1.2. Make sure a suitable lamp or resistor is in series with the wire so as to limit the current from the power supply. Use a wooden stand (or any non-ferrous object) to support the card so it is horizontal.

2. Use the plotting compass to plot magnetic field lines near the wire as explained in Topic 10.2. You should find that the field lines are circles (in the plane of the card) centred on the wire.

Use the arrangement shown to observe the effect of:

Reversing the current: You should find that the direction of the plotting compass reverses. This shows that the magnetic field lines reverse direction when the direction of the current is reversed.

Moving the plotting compass away from the wire: You should find it points more towards 'magnetic north'. This is because the magnetic field due to the wire decreases in strength further from the wire so the Earth's magnetic field has more effect. Remember that the direction of a magnetic field line at any point is the direction of the force on the N pole of a magnet at that point.

The magnetic field of a current-carrying solenoid

A solenoid is a long coil of wire. When there is a current in it, there is a magnetic field in and around the solenoid. The magnetic field pattern is shown in Figure 14.1.3. Reversing the direction of current reverses the direction of the magnetic field lines. Increasing the current increases the strength of the field everywhere in the field.

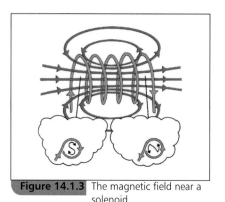

Figure 14.1.3 The magnetic field near a solenoid

We can use iron filings or a plotting compass to plot the field lines. Notice that each field line is a continuous loop through the inside of the solenoid and around the outside. The field lines are:
- concentrated at the ends of the solenoid
- parallel to each other inside the solenoid
- spread out beyond the ends of the solenoid.

Outside the solenoid, the magnetic field pattern is like that of a bar magnet. Using this comparison, we can say that the end of the solenoid from which the field lines emerge is like the north pole of a bar magnet and the other end is like the south pole. The end-view diagrams in Figure 14.1.3 show the 'solenoid rule', which is a simple way to relate the polarity of each end to the current direction.

Inside the solenoid, the magnetic field lines are parallel to each other so the field is uniform except near the ends. This means the magnetic field has the same strength and direction everywhere inside the solenoid (except near the ends where it is weaker).

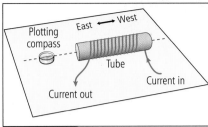

Figure 14.1.4 Investigating a solenoid

Supplement

PRACTICAL

Investigating the magnetic field of a solenoid

1   Make a solenoid using insulated wire wrapped around a cardboard or plastic tube. Fix the solenoid with its axis horizontal and aligned along an east-west line, as shown in Figure 14.1.4.

2   Place a plotting compass near one end of the solenoid then switch the solenoid current on. You should observe that the plotting compass points along the axis of the solenoid (instead of pointing to magnetic north when the current is off).

3   Reverse the current direction and you should see the plotting compass direction reverse.

4   Move the plotting compass through the inside of the solenoid. You should find the direction of the plotting compass is unchanged as it is moved along.

5   With the plotting compass on the solenoid axis just outside one end, move the plotting compass away from the solenoid along the axis. You should observe the plotting compass points more and more towards magnetic north as it is moved away from the solenoid. This shows that the strength of the magnetic field decreases with increased distance from the solenoid.

Applications

1 The uniform magnetic field inside a solenoid is used in applications such as the 'magnetic resonance' (MR) brain scanner.

2 The magnetic field outside a solenoid is used in many applications from large electromagnets used in scrap yards (see Topic 10.3) to small electromagnets used in relays (see Topic 13.6). Their iron core makes the magnetic field much stronger.

KEY POINTS

1   The magnetic field lines around a wire are circles centred on the wire in a plane perpendicular to the wire.

2   The magnetic field of a solenoid is uniform inside the solenoid and like that of a bar magnet outside.

3   Increasing the current increases the strength of the magnetic fields above; reversing the current reverses the magnetic field lines.

Supplement

SUMMARY QUESTIONS

1   Sketch the pattern of the magnetic field lines for each of the following:
    a   a vertical wire carrying current upwards,
    b   an air-filled solenoid.

2   A bar magnet is held near the end of an unfilled solenoid to repel it. Describe and explain the effect of:
    a   increasing the current in the solenoid,
    b   switching the current off,
    c   reversing the current,
    d   reversing the bar magnet.

The motor effect

Supplement

LEARNING OUTCOMES

- Describe the motor effect
- Relate the force in the motor effect to the current and the magnetic field

- Describe an experiment to show the effect of a magnetic field on an electron beam

Magnetic fields at work

How often do you use an electric motor? If you think you rarely use an electric motor, think again. Electric motors are used in many devices you use every day including computer disc drives, air-conditioning systems, fridges and electrically operated doors and windows.

An electric motor works because a force can be exerted on a wire in a magnetic field when a current passes through the wire. This effect, shown in Figure 14.2.1, is known as the **motor effect**. In addition to electric motors, the motor effect is made use of in other devices such as the loudspeaker and the analogue meter (see Topic 12.2). In the motor effect, the wire and the magnet (or electromagnet) exert equal and opposite forces on each other because their magnetic fields interact.

PRACTICAL

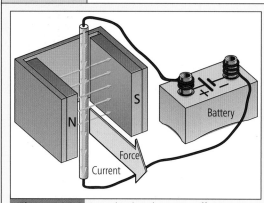

Figure 14.2.1 Investigating the motor effect

Investigating the motor effect

1 Figure 14.2.1 shows one way to investigate the motor effect. You should find that a force acts on the wire unless it is parallel to the magnetic field lines. Show that:

- increasing the current (by using two cells) increases the force on the wire
- reversing the current (by reversing the cell in the circuit) reverses the direction of the force on the wire.

2 In addition, test the effect with the wire in different directions in the magnetic field. As explained below, you should find this affects the force in a big way!

STUDY TIP

You must learn these factors that affect the size and direction of the force.

Supplement

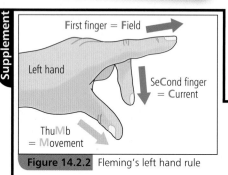

Figure 14.2.2 Fleming's left hand rule

Force factors

Your investigations should show that:

1 the force can be increased by:

- increasing the current
- using a stronger magnet.

2 the direction of the force is reversed if the direction of either the current or the magnetic field is reversed. If the current and the magnetic field are both reversed, the direction of the force is unchanged.

3 the force depends on the angle between the wire and the magnetic field lines. The force is:

- greatest when the wire is perpendicular to the magnetic field
- zero when the wire is parallel to the magnetic field lines.

4 the direction of the force is always at right angles to the wire and the field lines. The direction of the force can be worked out from the direction of the current and the direction of the magnetic field using Fleming's left hand rule shown in Figure 14.2.2.

The combined magnetic field

Figure 14.2.3 shows the pattern of the magnetic field lines of a current-carrying wire in a magnetic field. The magnetic field of the magnet and that of the wire cancel each other out on one side of the wire and reinforce each other on the other side. The result is that wire is pushed sideways by the magnet towards the weakest part of the field.

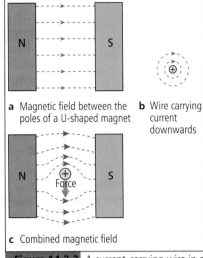

a Magnetic field between the poles of a U-shaped magnet   **b** Wire carrying current downwards

c Combined magnetic field

Figure 14.2.3 A current-carrying wire in a magnetic field of a U-shaped magnet

Supplement

The effect of a magnetic field on an electron beam

A magnetic field applied to a beam of electrons pushes the beam sideways, as shown in Figure 14.2.4. As the magnet is brought near the beam, the beam is pushed downwards because each electron in it experiences a downward force due to the magnetic field. The same happens to electrons moving along a current-carrying wire. The force on the wire and hence the motor effect is because the moving electrons are pushed sideways by the field.

In Figure 14.2.4: **the force on each electron is at right angles to the direction of the magnetic field and to the direction in which the electron is moving.**

If the magnet is turned around, the magnetic field direction is reversed so the beam is pushed upwards instead of downwards.

Using a current-carrying electromagnet instead of a permanent magnet would have the same effect. Increasing the current in the electromagnet increases the magnetic field strength of the electromagnet which increases the force on the electron beam and pushes it down further. Fleming's left hand rule gives the direction of the force on each electron in the beam provided the direction of the conventional current is used. See Topic 11.4.

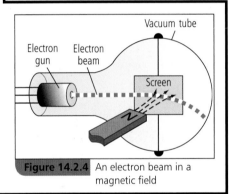

Figure 14.2.4 An electron beam in a magnetic field

SUMMARY QUESTIONS

1 Copy and complete the following sentences using words from the list below.

**horizontal   reversed   perpendicular
unchanged   vertical**

A vertical wire carrying an electric current is in a uniform horizontal magnetic field.

a   The direction of the force on the wire is _____ and _____ to the direction of the magnetic field lines.

b   When the current is increased, the force on the wire is _____.

c   When the magnetic field is reversed, the direction of the force on the wire is _____.

d   When the current in the wire is reversed and the magnetic field is reversed, the direction of the force on the wire is

_____.

2 In Figure 14.2.1, describe how the force on the wire changes if the wire is gradually turned until it is parallel to the magnetic field.

KEY POINTS

1 In the motor effect, the force:

- is increased if the current or the strength of the magnetic field is increased

- is reversed if the direction of the current or the magnetic field is reversed.

- is at right angles to the direction of the magnetic field and to the wire

2 Electrons in a beam are pushed sideways when a magnetic field is applied at right angles to the beam.

Supplement

The electric motor

Supplement

LEARNING OUTCOMES

- Explain why a current-carrying coil in a magnetic field turns
- Describe how an electric motor works
- Describe the effect of increasing or reversing the current in an electric motor or increasing the strength of the magnetic field

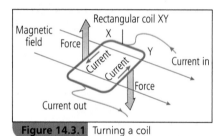

Figure 14.3.1 Turning a coil

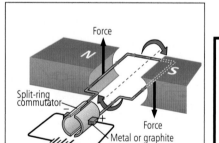

Figure 14.3.2 The electric motor

STUDY TIP

Study the diagrams with care and do your best to understand how the motor works.

Using the motor effect

A coil in a magnetic field can be forced to turn by passing a current through it. Figure 14.3.1 shows a rectangular coil between the poles of a U-shaped magnet. When a current passes around the coil, a force acts as shown on each of the two long sides of the coil. As these forces are in opposite directions, they act to make the coil turn.

PRACTICAL

Test a coil

1 Make a rectangular coil to fit between the poles of a U-shaped magnet. Use suitable insulated wire and a wooden block. Tape the wires to the block.

2 Place the coil in the field of a U-shaped magnet as shown in Figure 14.3.1. Use a suitable low voltage supply to pass a current through the wires of the coil. You should observe that the coil tries to turn.

  - Increase the current and you should observe the turning effect increases.
  - Reverse the current and you should observe the coil tries to turn in the opposite direction.
  - Use a stronger magnet and you should observe that the turning effect increases.

The simple electric motor

Supplement

An electric motor is designed to use the motor effect.

The simple motor in Figure 14.3.2 consists of a rectangular coil of insulated wire (the armature coil) that is forced to rotate. The coil is connected via two metal or graphite 'brushes' to the battery. The brushes press onto a metal **'split ring' commutator** fixed to the coil.

When a current is passed through the coil, the coil spins because:

- a force acts on each side of the coil due to the motor effect
- the force on one side is in the opposite direction to the force on the other side.

The split ring commutator reverses the current around the coil every half-turn of the coil. Because the sides also swap over each half-turn, the coil is pushed in the same direction every half-turn.

We can control the speed of an electric motor by changing the current. Also, we can reverse its turning direction by reversing the current.

Make and test an electric motor

Make and test a simple electric motor like the one in Figure 14.3.2.

1   The coil could be wound on a wooden block with a narrow hollow tube (not shown) through it. With a thin rod through the tube as the spindle, the wooden block and the tube should be able to spin freely.

2   To make the coil, wrap 'single-core' insulated wire around the sides of the coil so that the two bare ends of the wire can be laid on insulating tape to either side of one end of the tube.

3   Bend two small pieces of metal foil so they fit on the tube over the bare ends of the wire to form the split-ring commutator, as shown in Figure 14.3.2. Use two narrow strips of tape or small elastic bands to hold the metal foil in place.

4   Use two suitable metal pins fixed to a board as the pivots to hold the coil in place between opposite poles of a U-shaped magnet. If necessary, make each pivot using a paper clip.

Use the bare ends of the two wires from the battery as 'brushes'. With practice, when these are held in contact with opposite sides of the split-ring commutator, the coil should spin.

Practical electric motors

In a practical electric motor, the rotating part of the motor or 'armature' consists of several evenly-spaced coils wound on an iron core as shown in Figure 14.3.3a. The iron core makes the field much stronger so the turning effect is much greater.

- Each coil is connected to its own section of the commutator. The result is that each coil in sequence experiences a turning effect when it is connected to the voltage supply, so the armature is repeatedly pushed around.

- Because the iron core makes the field radial as shown in Figure 14.3.3b, each coil is in the field for most of the time. As a result, the overall turning effect is much steadier than in a simple electric motor with one coil only, so the motor runs more smoothly.

DID YOU KNOW?

Graphite is a form of carbon which conducts electricity and is very slippery. It therefore causes very little friction when it is in contact with the rotating commutator.

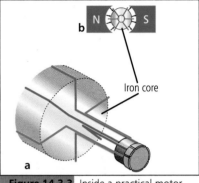

Figure 14.3.3   Inside a practical motor

KEY POINTS

1   A simple electric motor has a rectangular coil of wire that spins in a magnetic field when a current passes through the coil.

2   The speed of an electric motor is:

- increased if the current or the strength of the magnetic field is increased

- reversed if the current is reversed.

Supplement

1   Copy and complete the following sentences using words from the list below.

   coil   current   force   magnet

   a   When a _____ passes through the _____ of an electric motor, a _____ due to the _____ acts on each side of the _____.

   b   The _____ along each side is in opposite directions so the _____ is in opposite directions and the _____ turns.

2 a   Explain why a simple electric motor connected to a battery reverses if the battery connections are reversed.

   b   Discuss whether or not an electric motor would run faster if the coil was wound on:
   i   a plastic block,
   ii an iron block instead of a wooden block.

Electromagnetic induction

LEARNING OUTCOMES

- Describe how to induce an emf in a circuit
- State the factors that affect the magnitude of the induced emf
- Recognise that the direction of an induced current opposes the change that causes it

STUDY TIP

In a motor a current is supplied to a coil that is in a magnetic field. This produces the movement. In a generator the movement is supplied to a coil in a magnetic field and this produces the current

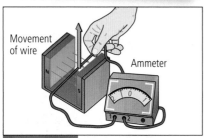

Movement of wire

Ammeter

Figure 14.4.1 The dynamo effect

STUDY TIP

Beware! Motors and generators both have a coil in a magnetic field. Many students see a diagram of one on the examination paper and write about the other!

A hospital has its own electricity generator always 'on standby' in case the mains electricity supply fails. Patients' lives would be put at risk if the mains power failed and there was no standby generator.

A generator contains coils of wire that spin in a magnetic field. A potential difference (pd) is created or **induced** in the wire when it cuts across the magnetic field lines. We refer to this source of pd as an **induced electromotive force** (emf) or an induced voltage. If the wire is part of a complete circuit, the induced emf makes an electric current pass around the circuit.

PRACTICAL

Investigating a simple generator

1 Connect some insulated wire to an ammeter as shown in Figure 14.4.1. Move the wire between the poles of a U-shaped magnet and observe the ammeter. You should discover the ammeter pointer deflects as a current is generated when the wire cuts across the magnetic field. The effect is known as the **dynamo effect**. The current is generated because a pd is induced in the wire when it cuts across the magnetic field lines. Make the wire into a coil and you should find the current is much greater.

2 Carry out tests to see what difference is made by:

- holding the wire stationary in the magnetic field,
- moving the magnet instead of the wire,
- moving the wire faster across the magnetic field,
- reversing the direction of motion of the wire.

You should find that:

- there is no current generated when the wire is stationary,
- a current is generated when the magnet instead of the wire is moved,
- a larger current is generated when the wire moves faster,
- the current is reversed when the direction of motion is reversed.

A generator test

Figure 14.4.2 shows a coil of insulated wire connected to a centre-reading ammeter.

When one end of a bar magnet is pushed into the coil, the ammeter pointer deflects. This is because:

- the movement of the bar magnet causes an induced emf in the coil
- the induced pd causes a current because the coil is part of a complete circuit.

In Figure 14.4.2, if the bar magnet is then withdrawn from the coil, the ammeter pointer deflects in the opposite direction. This is because the induced emf acts in the opposite direction so the induced current is in the opposite direction. The direction of the induced current also depends on which way around the polarity of the magnet is. For example, in Figure 14.4.2, the north pole of the bar magnet is shown entering end X of the coil.

The table below shows the results of testing each direction of motion of the magnet with the magnet each way around. The table gives the current direction as seen by someone viewing end X of the coil.

magnetic pole entering or leaving the coil	pushed in or pulled out	current direction	induced polarity of X	magnet and coil
north pole	in	anticlockwise	north pole	repel
north pole	out	clockwise	south pole	attract
south pole	in	clockwise	south pole	attract
south pole	out	anticlockwise	north pole	repel

The induced current passing through the wires creates a magnetic field as long as the coil is moving. We can use the solenoid rule shown in Figure 14.4.4 to work out whether end X of the coil is like the north or the south pole of a bar magnet. The result of using the solenoid rule is shown in the 'induced polarity' column of the table. These results show that the induced polarity of end X always acts against the change that causes it. The electrical energy generated is the result of work done by the person moving the magnet to overcome the magnetic field of the induced current.

The induced current always acts in such a direction as to oppose the change that causes it.

When a wire cuts across the lines of force of a magnetic field, as in Figure 14.4.1, the direction of the induced current can be worked out using Fleming's right hand rule, shown in Figure 14.4.4.

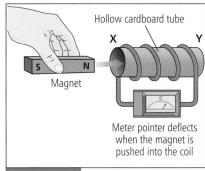

Figure 14.4.2 Testing electromagnetic induction

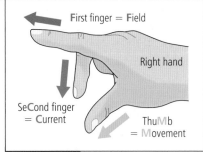

Figure 14.4.3 Fleming's right hand rule for induced current

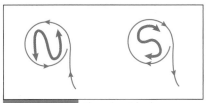

Figure 14.4.4 The solenoid rule

SUMMARY QUESTIONS

1 A coil of wire is connected to a centre-reading ammeter. A bar magnet is inserted into the coil, making the ammeter pointer flick briefly to the right.

   What would you observe:

   a if the magnet had been inserted more slowly into the coil?

   b if the magnet was then held at rest in the coil?

   c if the magnet is withdrawn rapidly from the coil?

2 a State two ways in which the voltage generated by a cycle dynamo could be increased.

   b Describe how you would use a coil, a bar magnet and an ammeter to demonstrate that the direction of an induced current opposes the change that causes it.

KEY POINTS

1 When a wire cuts the lines of a magnetic field, an emf is induced in a wire.

2 If the wire is part of a complete circuit, the induced emf causes a current in the circuit.

3 The current is increased if the wire moves faster or a stronger magnet is used.

4 The direction of an induced current opposes the change that causes it.

Supplement

LEARNING OUTCOMES

- Describe a simple alternating current generator
- Explain the use of the slip rings
- Sketch a graph to show how the induced emf of an ac generator varies with time
- Relate the position of the spinning coil in the magnetic field to the graph above

STUDY TIP

You might think that 'simple' is the wrong word here! Keep going, though; you can understand this!

The simple alternating current generator

A simple ac generator consists of a rectangular coil which is forced to spin in a magnetic field, as shown in Figure 14.5.1. The coil is connected to a centre-reading meter via metal 'brushes' that press on two metal slip rings. The slip rings and brushes provide a continuous connection between the coil and the meter.

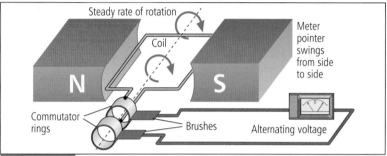

Figure 14.5.1 The construction of a simple ac generator

When the coil turns steadily in one direction, the meter pointer deflects first one way then the opposite way then back again. This carries on as long as the coil keeps turning in the same direction. The current in the circuit repeatedly changes its direction through the meter because the induced emf in the coil repeatedly changes its direction. This carries on as long as the coil keeps turning. The induced emf and the current are said to alternate because they repeatedly change direction.

The induced emf varies as the coil rotates, as shown in Figure 14.5.2. In one complete rotation of the coil (or 'one full cycle'), the induced emf increases from zero to a maximum value then decreases to zero, reverses and increases to a negative maximum and then becomes zero again. We refer to both the positive and negative maximum values as the **peak value**.

1 The magnitude of the induced emf is greatest when the plane of the coil is parallel to the direction of the magnetic field. At this position, the sides of the coil parallel to the axis of rotation (labelled X and Y in Figure 14.5.2) cut directly across the magnetic field lines. As a result, the induced emf is at its peak value.

2 The magnitude of the induced emf is zero when the plane of the coil is perpendicular to the magnetic field lines. At this position, the sides of the coil move parallel to the field lines and do not cut them. So the induced emf is zero.

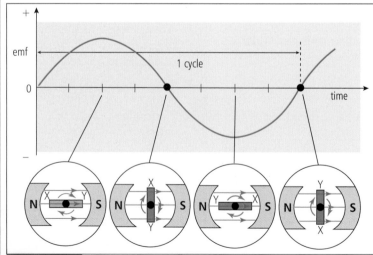

Figure 14.5.2 Alternating voltage

In one complete rotation of the coil starting with its plane **perpendicular** to the field, the induced emf therefore:

- increases from zero to a maximum as the coil turns through 90° to where its plane is parallel to the field
- decreases from the maximum to zero as the coil turns through a further 90° to where its plane is perpendicular to the field (and is upside down compared with its starting position)
- decreases from zero to a negative maximum as the coil turns through a further 90° to where its plane is once again parallel to the field (and is upside down compared with its starting position)
- increases from the negative maximum to zero as the coil turns through a further 90° to the position it was in at the start of the cycle.

Therefore each full cycle of the alternating emf takes the same time as one full rotation of the coil.

The faster the coil rotates:

1 **the greater the frequency** (i.e. the number of cycles per second) of the alternating current. This is because each full cycle of the alternating emf takes the same time as one full rotation of the coil.
2 **the larger the peak value** of the alternating current. This is because the sides of the coil move faster and therefore cut the field lines at a faster rate, so the induced emf is greater.

An alternating voltage can be displayed on an oscilloscope screen. If the induced emf from an ac generator is displayed on an oscilloscope and the generator is rotated faster, the screen display will show more waves on the screen (because the frequency of the induced emf will be greater). The waves will also be taller (because the peak value of the induced emf will be greater).

The cycle dynamo

Figure 14.5.3 shows the inside of a cycle dynamo. When the magnet spins, an alternating emf is induced in the coil. This happens because the magnetic field lines cut across the wires of the coil. The induced emf makes a current pass around the circuit when the lamp is on. Because the induced emf alternates, the current alternates too. The faster the magnet spins, the brighter the light is. This is because the induced emf is greater so a bigger current passes through the lamp.

STUDY TIP

Study this carefully along with the graph. Return to the waves section if you need to revise the meaning of 'frequency'.

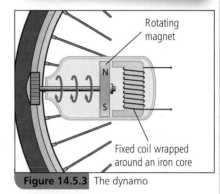
Figure 14.5.3 The dynamo

Rotating magnet

Fixed coil wrapped around an iron core

KEY POINTS

1 The simple ac generator consists of a coil that spins in a uniform magnetic field.

2 The slip rings and brush contacts enable the coil to stay connected to the external circuit.

3 The peak value of the induced emf is when the sides of the coil cut directly across the magnetic field lines.

4 When the sides of the coil move parallel to the field lines, the induced emf is zero.

SUMMARY QUESTIONS

1 Copy and complete the following sentences **a** to **c** using words from the list below:

alternates   reverses   spins   increases

a An emf is induced in the coil of an ac generator when the coil _____.

b The induced emf _____ which means its polarity repeatedly _____.

c If the coil _____ faster, the peak value of the induced emf _____ as well as its frequency.

2 Figure 14.5.2 shows how the alternating voltage produced by an ac generator changes with time.

a How would the graph in Figure 14.5.2 differ if the coil was rotated more slowly?

b Give reasons for your answer in **a**.

Supplement

LEARNING OUTCOMES

- Describe the construction of a transformer
- Recall and use the transformer equation relating the voltage ratio to the turns ratio
- Describe how a transformer works

DID YOU KNOW?

A mobile phone charger contains:

- a small transformer that steps the mains voltage down
- diodes that convert the alternating pd from the transformer to a direct voltage
- a capacitor to smooth out the direct voltage.

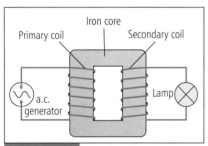

Figure 14.6.1 Transformer in action

STUDY TIP

Make sure you know the difference between a step-up and a step-down transformer and that you can draw the appropriate diagrams.

A typical power station generator produces an alternating voltage of about 25 000 V. The alternating voltage of the cables from a power station is typically 132 000 V. Mains electricity to homes and offices is much lower than this, for example 230 V in many countries.

Transformers are used to change an alternating voltage. We use them in low voltage supply units such as mobile phone chargers to step the alternating voltage from the mains down to the required lower voltage.

The construction of a basic transformer

A transformer has two coils of insulated wire, both wound around the same iron core as shown in Figure 14.6.1. The two coils are referred to as the primary coil and the secondary coil. When an alternating voltage from a suitable source is applied to the primary coil, an alternating voltage is induced in the secondary coil. If the secondary coil is part of a complete circuit, the alternating voltage induced in the secondary coil causes an induced current in the circuit containing the secondary coil.

The voltage applied to the primary coil must be an alternating voltage. A transformer will not work with a constant voltage applied to the primary coil. The alternating voltage causes an alternating current in the primary coil which induces an alternating voltage in the secondary coil. A constant voltage applied to the primary coil would cause a constant current in the primary coil and a constant current cannot cause electromagnetic induction.

A transformer is described as:

- **a step-up transformer** if the transformer 'steps up' the primary voltage so the secondary voltage is greater than the primary voltage
- **a step-down transformer** if the transformer 'steps down' the primary voltage so the secondary voltage is less than the primary voltage.

The transformer equation

In terms of peak values, the alternating voltage in the secondary coil, V_S, depends on:

1 the alternating voltage applied to the primary coil, V_P

2 the number of turns of wire in the primary coil, N_P

3 the number of turns of wire in the secondary coil, N_S.

The above quantities are related to each other through the following equation which we can use to calculate any one of these quantities if we know the other three:

$$\frac{\text{voltage across primary, } V_P}{\text{voltage across secondary, } V_S} = \frac{\text{number of turns on primary, } N_P}{\text{number of turns on secondary, } N_S}$$

- **For a step-up transformer**, the number of secondary turns N_s is greater than the number of primary turns N_p, so V_s is greater than V_p.
- **For a step-down transformer,** the number of secondary turns N_s is less than the number of primary turns N_p, so V_s is less than V_p.

Supplement

PRACTICAL

Make a model transformer

1 Wrap a coil of insulated wire around the iron core of a model transformer as the primary coil. Connect the coil to a 1 V ac supply unit and connect a second length of insulated wire to a 1.5 V torch lamp. When you wrap the second wire around the iron core, the lamp should light up.

2 Observe the effect of wrapping more turns of wire around the iron core. You should find the lamp is brighter. This is because the secondary voltage is greater.

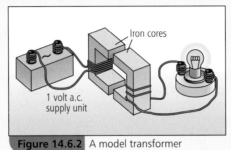

Iron cores

1 volt a.c. supply unit

Figure 14.6.2 A model transformer

WORKED EXAMPLE

A transformer is used to step a voltage of 230 V down to 10 V. The secondary coil has 60 turns. Calculate the number of turns of the primary coil.

Solution

$V_P = 230\,V$, $V_S = 10\,V$, $N_S = 60$ turns.

Using $\dfrac{V_P}{V_S} = \dfrac{N_P}{N_S}$

gives $\dfrac{230}{10} = \dfrac{N_P}{60}$

Therefore $N_P = \dfrac{230 \times 60}{10}$

$= 1380$ turns

KEY POINTS

1 A transformer consists of a primary coil and a secondary coil wrapped on the same iron core.

2 Transformers only work using alternating current.

The transformer equation is:

$$\frac{V_P}{V_S} = \frac{N_P}{N_S}$$

3 The alternating current in the primary coil creates an alternating magnetic field in the iron core which induces an alternating voltage in the secondary coil.

Supplement

How a transformer works

A transformer has two coils of insulated wire, both wound around the same iron core as shown in Figure 14.6.1. When alternating current is in the primary coil, an alternating voltage is induced in the secondary coil. This happens because:

- alternating current in the primary coil produces an alternating magnetic field
- the lines of the alternating magnetic field pass through the secondary coil and induce an alternating voltage in it.

If a lamp is connected across the secondary coil, the induced voltage causes a current in the secondary circuit. So the lamp lights up. Electrical energy is therefore transferred from the primary to the secondary coil. This happens even though they are not electrically connected in the same circuit.

SUMMARY QUESTIONS

1 Copy and complete the following sentences about a transformer using words from the list below.

current   voltage   primary   secondary

In a transformer, an alternating _____ is passed through the _____ coil. As a result, an alternating _____ is induced in the _____ coil.

2 a i Why does a transformer not work with direct current?

   ii Why is it important that the coil wires of a transformer are insulated?

 b A transformer with 1200 turns in the primary coil is used to step a voltage of 120 V down to 6 V. Calculate the number of turns on the secondary coil.

Supplement

LEARNING OUTCOMES

- Describe the use of transformers in the high-voltage transmission of electricity
- State the advantages of high voltage transmission
- Recall and use the power equation relating the current and pd in each coil for a transformer that is 100% efficient
- Explain why energy loss in cables is lower when the voltage is high

STUDY TIP

You must know that high voltage transmission is used because it is efficient – much less energy is wasted in the form of heat.

Electricity distribution

The electricity grid is the network of cables and transformers used to distribute mains electricity to our homes and other buildings from power stations.

The cables of the grid waste energy because the cables have resistance. Current in them has a heating effect. As a result, the cables do waste some energy. However, by operating the grid at a high voltage, we can reduce the current through the cables and transfer the same amount of electrical energy every second through them.

The higher the grid voltage, the greater the efficiency of transferring electrical power through the grid.

The efficiency may be expressed as the percentage of the electrical energy supplied to the grid that is used by the appliances and machines connected to it. The greater the efficiency, the lower the energy wasted in the cables. For example, if an electricity grid is supplied with 1000 million joules of electrical energy every second and 950 million joules is transferred to 'users', the efficiency of the grid is 95%. In other words, 5% of the electrical energy supplied every second to this grid is wasted.

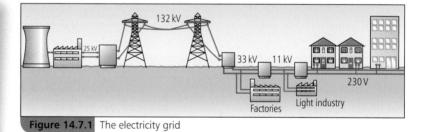

Figure 14.7.1 The electricity grid

Transformer efficiency

Supplement

Transformers are almost 100% efficient. This means that almost all the electrical energy supplied each second (i.e. electrical power) to the primary coil of a transformer is transferred to the device or devices connected to the secondary coil.

Since power = current × voltage:

- power supplied to the transformer = primary current, I_p × primary voltage, V_p
- power delivered by the transformer = secondary current, I_s × secondary voltage, V_s

Therefore, for 100% efficiency:

$$\text{primary current} \times \text{primary voltage} = \text{secondary current} \times \text{secondary voltage}$$

$$I_p V_p = I_s V_s$$

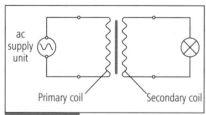

Figure 14.7.2 Transformer efficiency

WORKED EXAMPLE

A 120 V, 60 W lamp lights normally when it is connected to the secondary coil of a transformer and a 10 V ac supply is connected to the primary coil. The transformer is 100% efficient.

Calculate **a** the primary current, **b** the lamp current.

Solution

a Rearranging   power supplied to the primary coil

= the primary current × primary voltage,

$$\text{primary current} = \frac{\text{power supplied to the primary coil}}{\text{primary voltage}}$$

$$= \frac{60\,W}{10\,V} = 6.0\,A$$

b For 100% efficiency, power supplied to the lamp by the secondary coil = 60 W.

Rearranging   power supplied by the secondary coil

= the secondary current × secondary voltage,

$$\text{lamp current} = \frac{\text{power supplied to the lamp}}{\text{secondary voltage}} = \frac{60\,W}{120\,V} = 0.5\,A$$

Transformers and the grid

Transformers are used to step up the alternating voltage from a power station to the grid voltage and to step the grid voltage down to the mains voltage. The grid voltage is at least 132 000 V. What difference would it make if the grid voltage was much lower? Much more current would be needed to deliver the same amount of power. The grid cables would therefore heat up more and waste more power.

For example, to transfer 100 000 W of electrical power at 1000 V along a 0.5 Ω cable:

- the current passing along the cable would be 100 A (= 100 000 W/1000 V)
- the voltage drop along the cable would therefore be 50 V (= 100 A × 0.5 Ω).

So the power wasted in the cable would be 5000 W (= 100 A × 50 V). In other words, 5% of the power supplied to the cable is wasted as heat.

If the same amount of power was transferred along the same cable at 100 000 V instead of 1000 V:

- the current would be 1.0 A (= 100 000 W/100 000 V)
- the voltage drop along the cable would be 0.5 V (= 1.0 A × 0.5 Ω).

So the power wasted in the cable would be only 0.5 W (= 1.0 A × 0.5 V). In other words, the power wasted is negligible compared with the power transferred.

KEY POINTS

1 Transformers are used to step voltages up or down.

2 High voltage transmission of electricity is much more efficient than transmission at much lower voltages.

3 For a transformer that is 100% efficient:

the primary current × the primary voltage = the secondary current × secondary voltage

SUMMARY QUESTIONS

1 Copy and complete the following sentences using words from the list below:

down   primary secondary   up

a In a step-up transformer, the voltage across the _____ coil is greater than the voltage across the _____ coil.

b The voltage from a power station is stepped _____ so the same amount of power can be delivered through the cables as a result of stepping the current _____.

2 A transformer with a secondary coil of 100 turns is to be used to step a voltage down from 240 V to 12 V.

a Calculate the number of turns on the primary coil of this transformer.

b A 12 V, 36 W lamp is connected to the secondary coil. Calculate the current in **i** the lamp, **ii** the primary coil. Assume the transformer is 100% efficient.

1 (a) Describe an experiment to plot the shape of the magnetic field near a straight current-carrying wire. Include a diagram in your answer to show the apparatus used and to show the shape of the magnetic field produced.

(b) How would you find the direction of the magnetic field?

(c) What is the effect of reversing the direction of the current?

2 (a) Describe an experiment to plot the shape of the magnetic field in and around a long current-carrying coil (solenoid). Include a diagram in your answer to show the apparatus used and to show the shape of the magnetic field produced.

(b) How would you find the direction of the magnetic field?

(c) What is the effect of reversing the direction of the current?

3 Explain, in simple terms, the difference between an electric motor and a generator.

4 (a) Describe the construction of a basic step-up transformer. Include a diagram in your answer.

(b) Explain how the transformer works.

(c) Suggest a use for a step-up transformer.

5 A transformer is used to convert 240 V ac to 12 V ac. The primary coil has 48 000 turns.

(a) Calculate the number of turns on the secondary coil.

(b) State whether this is a step-up or step-down transformer.

(c) A 12 V, 5.0 A lamp is connected to the secondary coil. Calculate the current in the primary coil. Assume the transformer is 100% efficient.

1 The core of a transformer is made from

A   copper   B   iron   C   plastic   D   steel

(Paper 1/2)

2 The diagram shows a transformer.

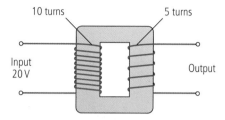

10 turns   5 turns

Input
20 V   Output

The output voltage is

A   2 V   B   10 V   C   20 V   D   40 V

(Paper 1/2)

3 The main advantage of high voltage transmission of electricity is

A   the current is high

B   the total resistance of the cables is high

C   the energy losses are low

D   the supply frequency is high.

(Paper 1/2)

4 In a generator, the size of the induced emf is greatest when

A   the plane of the coil is at right angles to the magnetic field

B   the plane of the coil is parallel to the magnetic field

C   the plane of the coil is at 45° to the magnetic field

D   the plane of the coil is at 30° to the magnetic field.

(Paper 2)

5 The diagram shows the magnetic field pattern around a straight current-carrying wire.

Plotting compass P has been correctly drawn to show the direction of the magnetic field.

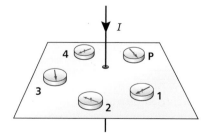

The other correctly drawn plotting compasses are

A 1 only   **C** 1, 2 and 3 only

B 1 and 2 only   **D** 1, 2, 3 and 4

(Paper 1/2)

6 The diagram shows a simple electric motor.

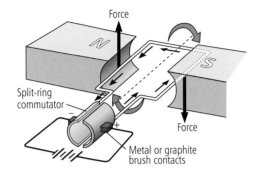

(a) Copy and complete the sentences to describe the working of an electric motor.

   (i) When there is a _____ in the coil, a _____ acts on each side of the coil.

  (ii) This is because the coil is in a _____ _____ .

 (iii) The split-ring commutator _____ the current every half turn so the motor can keep turning. *[4]*

(b) State two factors that would increase the speed of rotation of the motor. *[2]*

(c) State how you could reverse the direction of rotation of the motor. *[1]*

(Paper 3)

7 The diagram shows a beam of electrons in an electron tube. The electrons are emitted from an electron gun at the end of the tube. They travel down the tube at the same speed and hit a fluorescent screen at the other end of the tube, creating a spot of light at the point of impact.

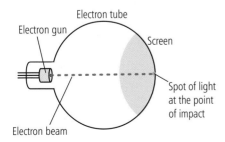

(a) When a bar magnet is held near the tube, the spot on the screen moves upwards. Explain why the spot moves upwards. *[2]*

(b) The magnet is removed and a small coil of wire attached to a battery and a switch is held near the tube as shown below.

   (i) When the switch is closed, the spot moved upwards again but not as far as before. Explain why the spot moves upwards but not as far this time. *[3]*

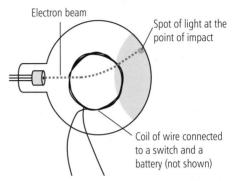

  (ii) State and explain what would be observed if the coil connections to the battery were reversed. *[2]*

(Paper 4)

8 The diagram shows a transformer.

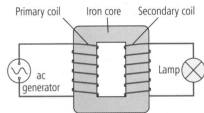

(a) Describe how the alternating current in the primary coil causes the lamp to light. *[4]*

(b) The diagram shows the same number of coils on the primary coil as on the secondary coil. Explain how this would be different if the operating voltage of the lamp were 1/10th of the supply voltage. *[2]*

(c) **(i)** A transformer is designed to provide a 6 V output from a 240 V input. The number of turns on the primary coil is 1200.

     Calculate the number of turns required on the secondary coil, assuming that the transformer is 100% efficient.

  (ii) The current in the secondary coil is 1 A. Assuming that the transformer is 100% efficient, calculate the current in the primary coil. *[5]*

(Paper 4)

15 Radioactivity

15.1

Observing nuclear radiation

LEARNING OUTCOMES

- State that radioactive substances emit radiation all the time
- State the three different types of radiation emitted by radioactive substances
- Describe how the different types of radiation from radioactive substances can be detected

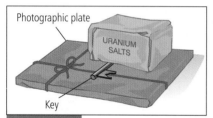

Figure 15.1.1 Becquerel's key

STUDY TIP

The word radiation is used in other contexts in Physics. Nuclear radiation has a particular meaning. Make sure you come to understand this as you work through this chapter.

A key discovery

If your photos showed a mysterious image, what would you think? In 1896, the French physicist **Henri Becquerel** discovered the image of a key on a film he developed. He remembered the film had been in a drawer under a key – with a packet of uranium salts on top. The uranium salts must have sent out some form of radiation that passed through paper (i.e. the film wrapper) but not through metal (i.e. the key).

Becquerel asked a young research worker, **Marie Curie,** to investigate. She found that the salts gave out radiation all the time. It happened no matter what was done to them. She used the word **radioactivity** to describe this strange new property of uranium. She and her husband, Pierre, did more research into this new branch of science. They discovered new radioactive elements. They named one of the elements polonium, after Marie's native country, Poland.

MARIE CURIE (1867–1934)

Becquerel and the Curies were awarded the Nobel prize for the discovery of radioactivity. Pierre died in a road accident. Marie went on with their work. She was awarded a second Nobel prize in 1911 for the discovery of polonium and radium. She died in middle-age from leukaemia in 1934. This is a disease of the blood cells and was caused by the radioactive materials she worked with.

Marie Curie (1867–1934)

PRACTICAL

Investigating radioactivity

We can use a **Geiger counter** to detect radioactivity. Look at Figure 15.1.2. The counter clicks each time a particle of radiation from a radioactive substance enters the Geiger tube.

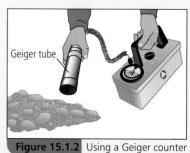

Figure 15.1.2 Using a Geiger counter

What stops the radiation? Ernest Rutherford carried out tests to answer this question about a century ago. He put different materials between the radioactive substance and a 'detector'. He discovered two types of radiation.

- One type (**alpha radiation**; symbol α) was stopped by paper.
- The other type (**beta radiation**; symbol β) went through it.
- Scientists later discovered a third type, **gamma radiation** (symbol γ), even more penetrating than beta radiation.

Radioactivity around us

When we use a Geiger counter, it clicks even without a radioactive source near it. This is due to **background radiation** from radioactive substances found naturally all around us.

Background radiation is ionising radiation from space (cosmic rays), from devices such as X-ray tubes and from radioactive substances in the environment. As explained in the next topic, the radiation from radioactive substances is harmful. Some of these substances are present because of nuclear weapons testing and nuclear power stations. But most of it is from substances in the Earth. For example, radon gas is radioactive and is a product of the decay of uranium in the ground. Radiation from substances in the Earth varies with location.

A radioactive puzzle

Why are some substances radioactive? Every atom has a nucleus made up of protons and neutrons. Electrons move about in the space surrounding the nucleus.

Most atoms each have a stable nucleus that doesn't change. But the atoms of a radioactive substance each have a nucleus that is unstable. An unstable nucleus becomes stable by emitting alpha, beta or gamma radiation. We say an unstable nucleus **decays** when it emits radiation. We can't tell when an unstable nucleus will decay. It is a **random** event that happens without anything being done to the nucleus.

DID YOU KNOW?

Radium is a radioactive element. It is so radioactive that it glows with light on its own. Many years ago, the workers in a US factory used radium paint on clocks and aircraft dials. They often licked the brushes to make a fine point to paint the dials accurately. Many of them later died from cancer.

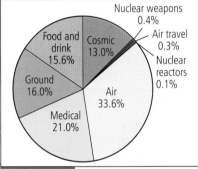

Figure 15.1.3 The origins of background radioactivity

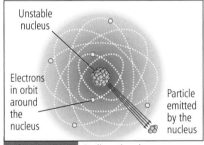

Figure 15.1.4 Radioactive decay

KEY POINTS

1 Radioactive substances give out radiation all the time.

2 Radioactive decay is a random event – we cannot predict or influence when it will happen.

3 Background radiation occurs naturally from traces of radioactive substances in buildings, in the atmosphere and in the ground.

4 There are three types of radiation from radioactive substances: alpha, beta and gamma radiation.

SUMMARY QUESTIONS

1 a The radiation from a radioactive source is stopped by paper. What type of radiation does the source emit?

b The radiation from a different source goes through paper. What can you say about this radiation?

2 Look at the pie chart in Figure 15.1.3.

a What is the biggest source of background radioactivity?

b List the sources in the chart and say which ones could be avoided.

Supplement

LEARNING OUTCOMES

- Describe the nature and main properties of α-, β- and γ-radiation

- Explain the relative ionising effect of α-, β- and γ-radiation

- Describe the deflection of α-, β- and γ-radiation using an electric field and using a magnetic field

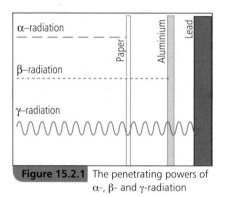

Figure 15.2.1 The penetrating powers of α-, β- and γ-radiation

STUDY TIP

The electrons that make up beta radiation come from the nucleus. They are NOT the electrons that orbit the nucleus.

Figure 15.2.4 Radioactive warning

Investigating the properties of alpha, beta and gamma radiation

Alpha radiation cannot penetrate paper. What stops beta and gamma radiation? How far can each type of radiation travel through air? The properties of each type of radiation can be demonstrated using a Geiger counter.

1 To test different materials, each type of material is placed between the tube and the radioactive source. We can add more layers of material until the radiation is stopped.

2 To test the range in air, the Geiger tube is moved away from the source. When the tube is beyond the range of the radiation, it cannot detect it.

The table below shows the results of the above tests.

type of radiation	absorber materials	range in air
alpha	paper	about 10 cm
beta	aluminium sheet (1 cm thick) lead sheet (2–3 mm thick)	about 1 m
gamma	thick lead sheet (several cm thick) concrete (more than 1 m thick)	unlimited see note below

Note: gamma radiation spreads out in air without being absorbed. It gets weaker as it spreads out.

The nature of alpha, beta and gamma radiation

What are these mysterious radiations? Experiments carried out by Ernest Rutherford and other scientists showed that:

- alpha radiation consists of positively charged particles, each consisting of two protons and two neutrons,

- beta radiation consists of electrons so it is negatively charged,

- gamma radiation is electromagnetic radiation so it is uncharged.

Deflection tests using electric and magnetic fields

Alpha, beta and gamma radiation in a narrow beam can be separated using a magnetic field, as shown in Figure 15.2.2.

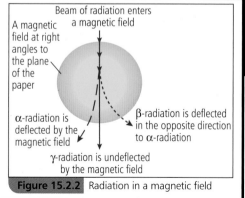

Figure 15.2.2 Radiation in a magnetic field

Supplement

1 β-radiation is easily deflected because it consists of electrons. See Topic 14.2.

2 α-radiation is deflected in the opposite direction to β-radiation because an α-particle has a positive charge. α-radiation is harder to deflect than β-radiation because an α-particle has much more mass than a β-particle, so much more force is needed to change its direction of motion.

3 γ-radiation is not deflected by a magnetic field. This is because gamma radiation is electromagnetic radiation so it is uncharged.

We can also use an electric field as shown in Figure 15.2.3 to deflect α- and β-radiation. Note that they are deflected in opposite directions.

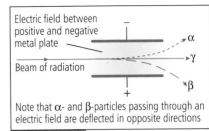

Electric field between positive and negative metal plate

Beam of radiation

Note that α- and β-particles passing through an electric field are deflected in opposite directions

Figure 15.2.3 Radiation in an electric field

Ionisation

The radiation from radioactive substances knocks electrons out of atoms. The atoms become charged as a result. The process is called **ionisation**. X-rays also cause ionisation. Alpha radiation has a much greater ionising effect than beta radiation, which has a much greater ionising effect than gamma radiation.

Ionisation in a living cell can damage or kill the cell. Damage to the genes in a cell can be passed on if the cell generates more cells. Strict rules must always be followed when radioactive substances are used.

Supplement

• Alpha radiation has a much greater ionising effect than beta radiation. This is because an α-particle has much more mass than a β-particle and moves more slowly, so it has much more effect on the atoms it encounters.

• Beta radiation has a greater effect than gamma radiation because β-particles are charged whereas gamma radiation is uncharged.

KEY POINTS

1 A radioactive substance contains unstable nuclei.

2 An unstable nucleus becomes stable by emitting radiation.

type of radiation	charge
alpha	positive
beta	negative
gamma	uncharged

SUMMARY QUESTIONS

1 Copy and complete the following sentences using words from the list:

   protons   neutrons   nucleus   radiation

   a The _____ of an atom is made up of _____ and _____.

   b When an unstable _____ decays, it emits _____.

2 a Copy and complete the following sentences using words from the list:

   alpha   beta   gamma.

   i Electromagnetic radiation from a radioactive substance is called _____ radiation.

   ii A thick metal plate will stop _____ and _____ radiation but not _____ radiation.

   b Which type of radiation is i uncharged ii positively charged iii negatively charged?

   c i Why is a radioactive source stored in a lead-lined box?

   ii Why should long-handled tongs be used to move a radioactive source?

3 Look at Figure 15.2.3:

   Explain why the α-particles are deflected upwards.

Supplement

The discovery of the nucleus

LEARNING OUTCOMES

- Describe Rutherford's alpha-particle scattering experiment and its results
- Explain how the results provide evidence for the nuclear model of the atom

Alpha particle scattering

Ernest Rutherford had already made important discoveries about radioactivity when he decided to use alpha particles to probe the atom. He asked two of his research workers, **Hans Geiger** and Ernest Marsden, to investigate the scattering of alpha particles by a thin metal foil. Figure 15.3.1 shows the arrangement they used.

The radioactive source they used had to decay slowly enough so its activity effectively stayed the same during the experiment. They measured the number of alpha particles deflected per second through different angles. The results showed that:

- most of the alpha particles passed straight through the metal foil,
- the number of alpha particles deflected per minute decreased as the angle of deflection increased,
- about 1 in 10 000 alpha particles were deflected by more than 90°,
- occasionally an alpha particle bounced off the foil back towards the source.

Rutherford was astonished by the results. He said it was like firing 'naval shells' at cardboard and discovering the occasional shell rebounds. He knew that alpha particles are positively charged. He deduced from the results that there is a **nucleus** at the centre of every atom. This nucleus is:

1 positively charged because it repels alpha particles (remember that particles with like charges repel),

2 much smaller than the atom because most alpha particles pass through without deflection,

3 where most of the mass of the atom is located.

Using this model, Rutherford worked out the proportion of alpha particles that would be deflected for a given angle. He found an exact agreement with Geiger and Marsden's measurements. He used his theory to estimate the diameter of the nucleus and found it was about 100 000 times smaller than the atom itself.

Rutherford's nuclear model of the atom was quickly accepted by other scientists because:

- it agreed exactly with Geiger and Marsden's measurements,
- it explains radioactivity in terms of changes that happen to an unstable nucleus when it emits radiation,
- it predicted the existence of the neutron, which was later discovered.

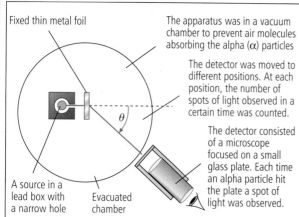

Figure 15.3.1 Alpha particle scattering

Fixed thin metal foil

The apparatus was in a vacuum chamber to prevent air molecules absorbing the alpha (α) particles

The detector was moved to different positions. At each position, the number of spots of light observed in a certain time was counted.

The detector consisted of a microscope focused on a small glass plate. Each time an alpha particle hit the plate a spot of light was observed.

A source in a lead box with a narrow hole

Evacuated chamber

θ

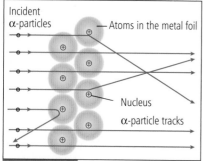

Figure 15.3.2 Alpha particle paths

Incident α-particles

Atoms in the metal foil

Nucleus

α-particle tracks

PRACTICAL

An α-scattering model

Fix a small metal disc about 2 cm thick at the centre of a table. Hide the disc under a cardboard disc about 20 cm in diameter. See if you can hit the metal disc with a rolling marble.

The scientist who split the atom!

When radioactivity was discovered, scientists couldn't work out what alpha radiation was – until Ernest Rutherford found the answer. He collected alpha particles in a tube. Then he made the 'alpha gas' light up by passing an electric current through it at high voltage. He found that light from the tube was the same as if helium gas was in the tube. So he concluded that alpha particles must be helium atoms without electrons.

Ernest Rutherford arrived in Britain from New Zealand in 1895. By the age of 28 he was a professor. He was awarded the Nobel Prize in 1908 for his investigations into radioactivity. He was knighted in 1914 for his discovery that every atom contains a nucleus. He went on to discover how to 'split the atom' and he found that nuclear reactions release much more energy then chemical reactions. He hoped no one would find out how to do this on a practical scale until the human race had learned to live in peace. He was made Lord Rutherford of Nelson in 1931. After his death in 1937, his ashes were placed close to Newton's tomb in Westminster Abbey in London.

Goodbye to the plum pudding atom!

Before the nucleus was discovered in 1914, scientists didn't know what the structure of the atom was. They did know it contained electrons and they knew these are tiny negatively charged particles. But they didn't know how the positive charge was arranged in an atom, although there were different models in circulation. Some scientists thought the atom was like a 'plum pudding' model with the positively charged matter in the atom evenly spread (like in a pudding) and the electrons buried inside (like plums in the pudding). Rutherford's discovery meant farewell to the plum pudding atom.

Lord Rutherford of Nelson (1871–1937)

DID YOU KNOW?

Imagine a marble at the centre of a football stadium. That's the scale of the nucleus inside the atom.

STUDY TIP

See Topic 4.4 Nuclear energy.

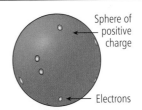

Figure 15.3.3 The plum pudding model

Sphere of positive charge

Electrons

KEY POINTS

1 Alpha particles in a beam are sometimes scattered through large angles when they are directed at a thin metal foil.

2 Rutherford used the measurements from alpha-scattering experiments to prove that an atom has a small positively charged central nucleus where most of the mass of the atom is located.

SUMMARY QUESTIONS

1 Copy and complete the following sentences using words from the list:

charge diameter mass

   a A nucleus has the same type of _____ as an alpha particle.

   b A nucleus has a much smaller _____ than the atom.

   c Most of the _____ of the atom is in the nucleus.

2 a Figure 15.3.4 shows four possible paths labelled A,B,C and D of an alpha particle deflected by a nucleus. Which path would the alpha particle travel along?

   b Explain why each of the other paths in a is not possible.

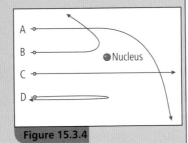

A

B ● Nucleus

C

D

Figure 15.3.4

Supplement

LEARNING OUTCOMES

- Recall and use the symbols for a nuclide
- Describe the changes to an unstable nucleus that take place due to a radioactive emission

STUDY TIP

Isotopes have the same number of protons but different numbers of neutrons.

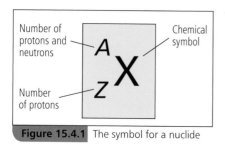

| Number of protons and neutrons | | Chemical symbol |

Figure 15.4.1 The symbol for a nuclide

Nuclides

The nucleus of an atom consists of protons which are positively charged and neutrons which are uncharged. Every atom of any particular element always has the same number of protons in its nucleus. However, the number of neutrons in the nucleus can differ. Each type of atom is called a **nuclide**. For example:

- Every atom of uranium has 92 protons in its nucleus. Most of these atoms contain 146 neutrons but some have 3 fewer neutrons.
- Every atom of carbon has 6 protons in its nucleus. Almost all these atoms contain 6 neutrons as well but some do not. For example, carbon atoms with 8 neutrons in the nucleus form a very tiny proportion of any sample of carbon.

Different nuclides with the same number of protons are the same element because every atom of an element has the same number of protons in its nucleus. The different nuclides of the same element are referred to as **isotopes** of the element.

The proton number (or atomic number), Z, of a nuclide is the number of protons in a nucleus of any atom of the element. The charge of a nucleus is therefore $Z \times$ the charge of a proton.

The nucleon number (or mass number), A, of a nuclide is the total number of protons and neutrons in any atom of that nuclide. As the mass of a neutron is almost the same as the mass of a proton, the mass of a nucleus is therefore almost equal to $A \times$ the mass of a proton (or a neutron).

Figure 15.4.1 shows the symbol for a nuclide. The values of A and Z always precede the chemical symbol with Z as the subscript. For example: for a uranium nuclide with 146 neutrons and 92 protons, $Z = 92$ and $A = 238$ $(= 92 + 146)$, thus the symbol is $^{238}_{92}U$

Radioactive changes

In general, each radioactive nuclide emits one type of radiation only which is either an α- or a β-particle or γ-radiation. When a nucleus emits an α- or a β-particle, the nucleus changes to that of a different element.

Supplement

Alpha radiation consists of particles, each composed of two protons and two neutrons. We use the symbol $^4_2\alpha$ for an α-particle, as $Z = 2$ (because it contains 2 protons) and $A = 4$ (because it consists of 2 protons and 2 neutrons).

When an α-particle is emitted from an unstable nucleus, the nucleus loses 2 protons and 2 neutrons, as shown in Figure 15.4.2. Therefore:

- the proton number of the nucleus decreases by 2 and its mass number decreases by 4
- it becomes a nuclide of a different element because it loses 2 protons.

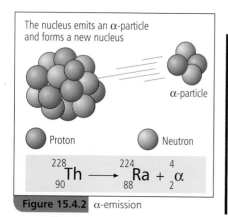

The nucleus emits an α-particle and forms a new nucleus

○ Proton   ○ Neutron

$$^{228}_{90}Th \longrightarrow {}^{224}_{88}Ra + {}^4_2\alpha$$

Figure 15.4.2 α-emission

We can represent this change by an equation like the one shown here:

$$^{228}_{90}\text{Th} \longrightarrow \ ^{224}_{88}\text{Ra} + \ ^{4}_{2}\alpha$$

Notice that the total number of protons and the total number of neutrons after the change is the same as before the change.

1 The total number of protons on each side are equal: $90 = 88 + 2$

2 The total number of protons and neutrons on each side are equal, i.e. $228 = 224 + 4$

Beta radiation consists of electrons. We use the symbol $^{0}_{-1}\beta$ for a β-particle where $Z = -1$ (as its charge is equal and opposite to that of a proton) and $A = 0$ (because it does not contain any protons or neutrons).

When a β-particle is emitted from an unstable nucleus, a neutron in the nucleus changes into a proton and a β-particle is created and emitted at the same time. Figure 15.4.3 shows the process. Therefore:

• the proton number of the nucleus increases by 1 and its mass number is unchanged because the total number of protons and neutrons is unchanged.

• it becomes a nuclide of a different element because it has one more proton and one less neutron than it started with.

We can represent this change by an equation like the one shown below:

$$^{40}_{19}\text{K} \longrightarrow \ ^{40}_{20}\text{Ca} + \ ^{0}_{-1}\beta$$

1 The nucleus has 19 protons and 21 ($=40-19$) neutrons before the change. After the change, it has 20 protons and 20 neutrons. A neutron has turned into a proton and a β-particle has been released.

2 The total number of protons and neutrons on each side are equal. There is a total of 40 protons and neutrons in the nucleus before the change and after the change.

Gamma radiation is high-energy electromagnetic radiation. After an unstable nucleus has emitted an α-particle or a β-particle, it sometimes has surplus energy. It emits this energy as γ-radiation.

STUDY TIP

Practise balancing nuclear equations until you are confident with them.

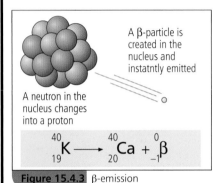

A β-particle is created in the nucleus and instatntly emitted

A neutron in the nucleus changes into a proton

$$^{40}_{19}\text{K} \longrightarrow \ ^{40}_{20}\text{Ca} + \ ^{0}_{-1}\beta$$

Figure 15.4.3 β-emission

SUMMARY QUESTIONS

1 Work out the number of protons and the number of neutrons in each of the following nuclides.

   a $^{228}_{90}\text{Th}$ **b** $^{234}_{91}\text{Pa}$ **c** $^{227}_{89}\text{Ac}$

2 **a** $^{238}_{92}\text{U}$ is an α-emitter.

   i When such a nucleus emits an α-particle, which one of the following nuclides is formed?

      $^{234}_{92}\text{U}$ $^{233}_{90}\text{Th}$ $^{234}_{90}\text{Th}$ $^{228}_{90}\text{Th}$ $^{231}_{90}\text{Th}$

   ii Write an equation for the above change.

  b $^{234}_{91}\text{Pa}$ is a β-emitter.

   i From the list of nuclides above in **ai**, identify the nucleus formed when this change happens.

   ii Write an equation for the above change.

KEY POINTS

1 A nuclide $^{A}_{Z}\text{X}$ contains Z protons and $(A - Z)$ neutrons.

2 When a nucleus emits:

• an α-particle, it loses 2 protons and 2 neutrons

• a β-particle, a neutron in the nucleus changes into a proton

• γ radiation, the number of neutrons and protons it contains is unchanged.

LEARNING OUTCOMES

- Explain what is meant by the term *half-life*
- Use half-life in calculations or in graphs
- Describe how radioactive materials are stored and used safely

Radioactive decay

The **activity** of a radioactive substance is the number of nuclei that decay per second. Each unstable nucleus (the 'parent' nucleus) changes into a nucleus of a different isotope (the 'daughter' nucleus) when it decays. Because the number of parent nuclei goes down, the activity of the sample decreases. We say the sample **decays**.

We can use a Geiger counter to monitor the activity of a radioactive sample. We need to measure the **count rate** due to the sample. This is the number of counts per second (or per minute) due to the sample only. The Geiger counter measures counts due to background radiation and due to the sample. So we need to subtract the background count rate from the measured count rate to obtain the count rate due to the sample.

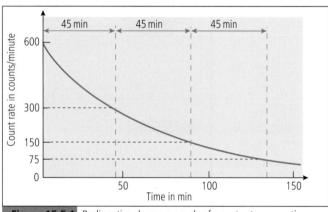

Figure 15.5.1 Radioactive decay: a graph of count rate versus time

Fig 15.5.1 shows how the count rate of a sample decreases with time. It falls from:

- 600 counts per minute (cpm) to 300 cpm in the first 45 minutes
- 300 counts per minute (cpm) to 150 cpm in the next 45 minutes.

The time taken for the count rate (and therefore the number of parent atoms) to fall by half is always the same. This time is called the **half-life**. The half-life in Figure 15.5.1 is 45 minutes.

The half-life of a radioactive substance is the time it takes:

- **for the number (and therefore the mass) of parent nuclei in a sample to halve**
- **for the count rate from the original substance to fall to half its initial level.**

The random nature of radioactive decay

We can't predict *when* an individual nucleus will suddenly decay. But we *can* predict how many nuclei will decay in a certain time – because there are so many of them. This is a bit like throwing dice. If you threw 1000 dice, you would expect one-sixth to land on a particular number.

Suppose we start with 1000 unstable nuclei, as in Figure 15.5.2.

If 10% disintegrate every hour, we would expect:

100 nuclei to decay in the first hour, leaving 900

90 nuclei (= 10% of 900) to decay in the second hour, leaving 810

81 nuclei (= 10% of 810) to decay in the second hour, leaving 729.

Supplement

If a graph is plotted without subtracting the background rate, the curve flattens out above zero at the level of the background rate. This level can be estimated from the graph to give the background rate. Subtracting this estimate from the measured count rate then gives the count rate due to the sample.

PRACTICAL

Modelling radioactive decay

Do an investigation modelling radioactive decay using dice or a computer simulation.

The table shows what you get if you continue the calculations.

time from start / hours, h	0	1	2	3	4	5	6	7
number of unstable nuclei present	1000	900	810	729	656	590	530	477
number of unstable nuclei that decay in the next hour	100	90	81	73	66	59	53	48

Using radioactive materials

Radioactive materials are used in nuclear power stations and in many hospitals and laboratories and in certain factories. When a radioactive source is not in use, it must be stored in a thick lead container in a secure location that is not accessible to unauthorised users. In use, to minimise exposure of users to ionising radiation, the radioactive source should be:

- kept at a safe distance from the user (or users),
- separated from users by thick lead screens if necessary,
- in use for the shortest possible time,
- only moved using long-handled tools so it is not near to anybody.

After use, it should be immediately returned to its container and its storage location.

Radiation workers are required to carry a radiation monitoring device, usually a film badge attached to clothing. The film badge is sealed from light and contains a strip of photographic film which is covered in different parts by different filters. Ionising radiation darkens the covered parts of the film by different amounts according to the filter in each part. In this way, the film can be used to find out the total exposure the wearer has been subjected to.

Unwanted radioactive substances, such as radioactive waste from nuclear power stations and unwanted radioactive sources from hospitals, factories, laboratories, schools and other users, must be disposed of safely and securely in strict accordance with legally-enforced safety regulations.

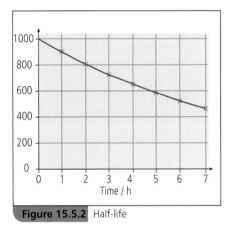

Figure 15.5.2 Half-life

A film badge

SUMMARY QUESTIONS

1 Copy and complete the following sentences using words from the list below.

half-life   stable   unstable

a In a radioactive substance _____ nuclei decay and become

_____.

b The _____ of a radioactive substance is the time taken for the number of _____ nuclei to decrease to half.

2 a A radioactive substance has a half-life of 15 h. A sealed tube contains 8 mg of the substance. What mass of the substance is in the tube **i** 15 h later **ii** 45 h later?

b In Figure 15.5.1 what will the count rate be after 135 min from the start?

c In Figure 15.5.2, use the graph to measure the half-life of this radioactive isotope.

KEY POINTS

1 The half-life of a radioactive substance is the time it takes for the number of parent nuclei in a sample to halve.

2 The number of unstable atoms and the activity decreases to half in one half-life.

Radioactivity at work

LEARNING OUTCOMES

- Define and use the term *isotope*
- Describe practical applications of radioactive isotopes
- Recognise in a practical application the significance of the half-life of a radioactive source and the type of radiation it emits

Isotopes are atoms with the same number of protons but different numbers of neutrons. We use radioactive isotopes for many purposes. For each purpose, we need a radioactive isotope that emits a certain type of radiation and has a suitable half-life.

1 Automatic thickness monitoring is used in the production of metal foil, as shown in Figure 15.6.1. A source that emits β-radiation is used. The amount of β-radiation passing through the foil depends on the thickness of the foil. A detector on the other side of the metal foil measures the amount of radiation passing through it.

- If the thickness of the foil increases too much, the detector reading drops.
- The detector sends a signal to the rollers to increase the pressure on the metal sheet. This makes the foil thinner.

The source needs to have a long half-life so its activity hardly changes while it is being used. This ensures variations in the count rate are due only to variations in the thickness of the foil.

The source needs to emit β-radiation so that variations in the foil thickness affect the amount of radiation passing through the foil and reaching the detector. Using a β-emitting source therefore ensures the reading of the detector decreases if the thickness of the metal foil increases, and increases if the thickness of the metal foil decreases.

A source that emits α-radiation would be unsuitable because α-radiation is stopped by thin metal foil so would not reach the detector. A source that emits γ-radiation would also be unsuitable because γ-radiation passes through metal foil regardless of its thickness.

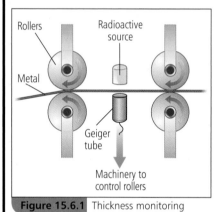

Figure 15.6.1 Thickness monitoring

STUDY TIP

There are many uses for radioactive isotopes. The suitability of an isotope depends on its half-life and the type of radiation emitted (which determines how penetrating the radiation is).

2 Radioactive tracers are used to trace the flow of a substance through a system. For example, doctors use radioactive iodine to find out if a patient's kidney is blocked.

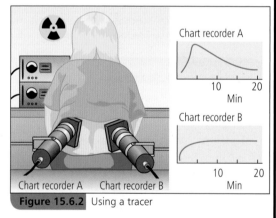

Figure 15.6.2 Using a tracer

Before the test, the patient drinks water containing a tiny amount of the radioactive substance. A detector is then placed against each kidney. Each detector is connected to a chart recorder as shown in Figure 15.6.2. The radioactive substance flows in and out of a normal kidney. So the detector reading goes up then down as shown by chart recorder A.

For a blocked kidney, the reading goes up and stays up as shown by chart recorder B. This is because the radioactive substance goes into the kidney but doesn't flow out again.

The radioactive iodine isotope $^{131}_{53}I$ is used for this test because:

- its half-life is 8 days so it lasts long enough for the test to be done and decays almost completely after a few weeks,
- it emits gamma radiation so it can be detected outside the body,
- it decays into a stable, safe product.

3 **Radioactive dating** is used to find how old ancient material is (its age).

- **Carbon dating** is used to find the age of ancient wood. The carbon content of living wood is made up mainly of the non-radioactive carbon isotope $^{12}_{6}C$ and a tiny proportion of the radioactive carbon isotope $^{14}_{6}C$. This has a half-life of 5600 years. When a tree dies, the amount of radioactive carbon in it decreases. To find the age of a sample, we need to measure the count rate from it. This is compared with the count rate from the same mass of living wood. For example, suppose the sample count rate is half the count rate of an equal mass of living wood. Then the age of the sample must be 5600 years.

- **Uranium dating** is used to find the age of igneous rocks. These rocks contain the uranium isotope $^{238}_{92}U$, which has a half-life of 4500 million years. Each uranium nucleus decays into a nucleus of lead which is stable. We can work out the age of a sample by measuring the number of atoms of uranium and lead. For example, if the sample contains 1 atom of lead for every atom of the uranium, the age of the sample must be 4500 million years. This is because there must have *originally* been 2 atoms of uranium for each atom of uranium now present.

DID YOU KNOW?

Smoke alarms save lives. A radioactive source inside the alarm sends out alpha particles into a gap in a 'detector' circuit in the alarm. The alpha particles ionise the air in the gap so it conducts a current across the gap. In a fire, smoke absorbs the ions created by the alpha particles so they don't cross the gap. The current across the gap drops and the alarm sounds. The battery in a smoke alarm needs to be checked regularly – to make sure it is still working.

A smoke alarm

SUMMARY QUESTIONS

1 Copy and complete the following sentences using words from the list below.

alpha   beta   gamma

a In the continuous production of thin metal sheets, a source of _____ radiation should be used to monitor the thickness of the sheets.

b A radioactive tracer given to a hospital patient needs to emit _____ or _____ radiation.

c The radioactive source used to trace a leak in an underground pipeline should be a source of _____ radiation.

2 a Why is alpha radiation not suitable for kidney tests?

b Why couldn't carbon dating be used to find how old a living tree is?

c What could you say about an igneous rock with uranium but no lead in it?

KEY POINTS

1 Isotopes are atoms with the same number of protons and a different number of neutrons in the nucleus.

2 The use we can make of a radioactive isotope depends on:

- its half-life
- the type of radiation it emits.

SUMMARY QUESTIONS

1 (a) Explain the meaning of the term *background radiation*.

(b) Suggest three sources of background radiation.

2 (a) State the nature of:
   (i) alpha radiation
   (ii) beta radiation
   (iii) gamma radiation.

(b) For each of the three types of radiation describe:
   (i) the charge (if any)
   (ii) the ionising properties
   (iii) the penetrating power
   (iv) the effect of an electric or magnetic field.

3 (a) Describe, with the aid of a diagram, an experiment to find the contributions of alpha, beta and gamma radiation coming from a small sample of radioactive material.

(b) List the safety precautions that you would take when carrying out the experiment.

4 (a) Define the term *half-life*.

(b) Define the term *isotope*.

(c) Explain what is meant by the penetrating power of a radioactive emission.

(d) Suggest two uses for radioactive isotopes and explain the importance of the half-life and penetrating power for each of the uses you choose.

Supplement

5 The diagram shows the apparatus used by Ernest Rutherford when using alpha particles to study the structure of the atom.

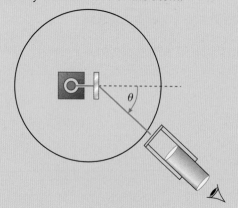

PRACTICE QUESTIONS

1 There are three types of radioactive emission: alpha, beta and gamma. How many of these types of emission is/are a stream of particles?

   A   0
   B   1
   C   2
   D   3

   (Paper 1/2)

2 A radioactive substance emits radiation that passes through paper, but is stopped by a thin sheet of steel. The type(s) of radiation emitted is/are

   A   alpha only
   B   alpha and beta only
   C   beta only
   D   gamma only

   (Paper 1/2)

3 A uranium atom has an atomic mass A of 238 and an atomic number Z of 92. The number of neutrons in the nucleus is

   A   92
   B   146
   C   238
   D   330

   (Paper 1/2)

4 The count rate for a radioactive sample is 3000 counts/minute. One hour later the count rate is 750 counts/minute. The half-life of the sample is

   A   15 min
   B   20 min
   C   30 min
   D   40 min

   (Paper 1/2)

(a) Describe the main findings about the paths of the alpha particles.

(b) Explain the conclusions that the experiment provided.

(c) How were the alpha particles detected?

5 The nuclear equation

$$^{14}_{6}C \longrightarrow \; ^{14}_{7}N + \ldots$$

shows that carbon-14 emits:

A alpha particles

B beta particles

C gamma radiation

D neutrons

(Paper 2)

6 (a) Copy and complete the sentences:

(i) Alpha radiation is a stream of particles each of which is composed of two _____ and two _____ .

(ii) Alpha particles have a charge of _____ .

(iii) Beta radiation is a stream of fast-moving _____ .

(iv) Beta particles have a charge of _____ .

(v) Gamma radiation is an _____ _____, it has _____ charge. *[7]*

(b) The half-life of a radioisotope is 4 min. The count rate is measured and found to be 5600 counts/minute. Calculate the count rate after 16 min. *[3]*

(Paper 3)

7 The diagram shows a radioactive source, a Geiger tube and counter and a holder for an absorber.

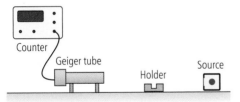

(a) A student first measures the average background radiation count. Explain what is meant by the term *background radiation*. *[1]*

(b) The source is thought to be a beta emitter. Describe how you could use the apparatus, with suitable absorbers, to show that the source emits beta particles but not alpha particles or gamma radiation. *[5]*

(c) Suggest two safety precautions that you would take while carrying out the experiment. *[2]*

(Paper 3)

8 (a) The diagram shows the paths of alpha particles passing through a very thin layer of gold.

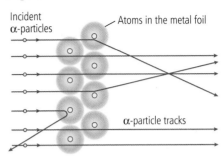

(i) Explain why most alpha particles pass through undeflected.

(ii) Explain why some alpha particles are deflected through small angles.

(iii) Explain why a very few alpha particles are deflected through an angle of more than 90°.

(iv) State the main conclusions from this experiment. *[7]*

(b) An old theory of the structure of the atom describes it as having a widely spread area of positive charge with tiny negative charges within it. Explain why this arrangement would not cause alpha particles to be deflected through large angles. *[2]*

(Paper 4)

Glossary

A

absolute scale of temperature temperature scale in kelvin (K) with a fixed point at absolute zero (0 K) such that $0\,°C = 273\,K$ and $100\,°C = 373\,K$

absolute zero the lowest possible temperature, 0 kelvin (K) or $-273\,°C$

ac generator a device consisting of a coil that spins in a magnetic field and generates an alternating emf as a result

acceleration change of velocity per second, measured in metres per second^2 (m/s^2)

accuracy of a measurement the difference between the measured reading and the true reading

alpha radiation radiation consisting of particles emitted by certain radioactive substances; an α-particle consists of 2 protons and 2 neutrons; α-particles are stopped by paper, deflected by an electric or magnetic field and are very ionising.

ammeter electrical instrument used to measure electric current in amperes

amplitude the maximum distance of any part of a wave from its undisturbed position; for a transverse wave, this is the height of a wave crest or the depth of a wave trough from the middle

analogue circuit circuits in which the voltage at any point can be at any value between the maximum and minimum values of the power supply (e.g. any value between $+5\,V$ or zero in a circuit with a $5\,V$ power supply)

analogue signal signal that varies and can have any value between a maximum and a minimum

angle of incidence angle between the incident ray and the normal

angle of reflection angle between the reflected ray and the normal

average speed total distance travelled ÷ total time taken, measured in metres per second (m/s)

B

background radiation alpha (α) or beta (β) or gamma (γ) radiation emitted by radioactive substances that occur naturally all around us

barometer instrument used to measure atmospheric pressure

beta radiation radiation consisting of electrons emitted by certain radioactive substances; a β-particle is emitted by an unstable nucleus when a neutron in the nucleus changes into a proton to enable the nucleus to become more stable; β-particles are stopped by several mm of aluminium paper, easily deflected by an electric or magnetic field and are ionising

boiling change of state from liquid to vapour at the boiling point

Boyle's Law for a fixed mass of gas at constant temperature, its pressure × its volume = constant

Brownian motion the random motion of particles in a fluid (such as smoke particles in air) due to collisions with individual fluid molecules

C

Celsius scale temperature scale based on ice point (the temperature of pure melting ice) defined as $0\,°C$ and steam point (the temperature of steam at standard atmospheric pressure) defined as $100\,°C$

centre of mass every object behaves as if all its mass were concentrated at one point called its centre of mass

centripetal force the force needed to keep an object moving around in a circle

charging by direct contact process in which an insulated conductor is charged by direct contact with a charged body

charging by induction process in which an insulated conductor is charged by earthing it briefly in the presence of a charged body; the insulated conductor gains the opposite type of charge to that of the charged body

chemical energy energy released when a chemical reaction occurs

circuit breaker an electromagnetically-operated switch that opens if too much current passes through it and which needs to be reset manually once it has opened.

compression in a wave part of a sound wave where air molecules are pushed together

condensation change of state from vapour to liquid at the boiling point or from vapour to solid directly

conduction electrons see free electrons

conduction of heat transfer of energy due to a temperature difference within a substance as a result of vibrations of atoms in any substance and the movement of free electrons in a metal

conservation of energy the total amount of energy before and after a change is the same

convection transfer of energy in a liquid or a gas due to a temperature difference causing circulation within the liquid or gas

converging lens a lens that focuses parallel rays of light to a point

critical angle the angle of incidence of a light ray in a transparent substance when it is refracted along the boundary

D

decelerates slows down

density mass per unit volume of an object, measured in kilograms per cubic metre (kg/m^3)

diffraction spreading of waves when they pass through a gap or move around an obstacle

diffusion process in which free electrons in a metal spread out from a region of high temperature to regions at lower temperature or particles in a substance spread out from a region of high concentration to surrounding regions of lesser concentration

digital circuit circuits in which the voltage at any point is either high, '1', (e.g. $+5\,V$) or low, '0', (e.g. zero)

digital signal signal consisting of digital pulses (i.e. 0's and 1's)

diode a component that allows current to pass through it in one direction only (its 'forward' direction)

dispersion splitting of white light (or any non-monochromatic light) into separate colours using a glass prism; the effect occurs because the refractive index of glass varies with the colour of light

drag force resistance to the motion of an object moving through air or any other fluid

E

earth wire a word used to describe a wire used to connect a metal object to the ground so the object cannot retain charge

echo reflection of sound from a smooth surface

efficiency useful energy transferred by a device ÷ energy supplied to the device

elastic limit the limit to which a material can be stretched without being permanently extended

elastic object an object that regains its shape after being distorted is said to be elastic

elastic strain energy energy stored in an object due to a change of its shape

electric charge there are two types of electric charge: positive and negative; charged objects attract or repel one another according to whether or not they have the same type of charge, in which case they repel, or they have different (i.e. unlike or opposite) types of charge, in which case they attract. Charge is measured in coulombs (C)

electric current a flow of charge; measured in amperes (A) where 1 ampere is a rate of flow of charge of 1 coulomb per second

electric field the space surrounding a charged object in which any other charged object experiences a force due to its charge

electric line of force a line in an electric field in which a small positively charged object would move if free to do so

electrical conductor a substance through which an electric current can flow

electrical energy energy transferred by an electric current

electrical insulator a substance through which an electric current cannot flow

electrical power electrical energy transferred per second, measured in watts (W). Electrical power = current × pd

electricity grid network of cables and transformers used to distribute mains electricity produced by power stations to buildings

electromagnet a device consisting of insulated wire wrapped around an iron core; when there is a current in the wire, the iron core is magnetic and will attract unmagnetised ferrous objects

electromagnetic induction the generation of an emf in a wire when the wire cuts across the lines of a magnetic field; the emf increases the faster the wire moves across the field lines and reverses if the field or the direction of motion of the wire is reversed

electromagnetic spectrum spectrum of electromagnetic waves; in order of increasing frequency (and decreasing wavelength): radio waves, microwaves, infra-red radiation, light, ultraviolet radiation, X radiation, gamma radiation

electromagnetic waves electric and magnetic disturbances that transfer energy from one place to another

electromotive force (emf) the 'push' provided by a cell or battery to force charge to flow around a circuit; it is measured in volts (V) where the volt is the energy that a cell or battery can deliver to the components in a circuit when a given amount of charge (i.e. 1 coulomb) flows around the circuit

electron a negatively charged particle which has a much smaller mass than the proton and orbits the nucleus in every atom

electroscope an instrument used to detect small amounts of charge

electrostatic precipitator an electrostatic device used to remove particles of ash and dust from the flue gases produced in a coal-fired power station

endoscope a medical instrument used to view objects and surfaces in cavities inside the body

energy, dissipation of energy tends to spread out, for example where energy is transferred to the surroundings by heating or by sound waves

energy transfers ways of storing and transferring energy

equilibrium an object at rest is said to be in equilibrium

evaporation the process in which a liquid turns to vapour below its boiling point

extension increase of the length of an object

F

ferrous material that contains iron

fission reactor nuclear reactor in which energy is released due to nuclear fission of uranium

fixed points standard 'degrees of hotness', usually melting or boiling points of pure substances, used to define a temperature scale

fluorescent substance substance that glows when high-energy radiation or particles such as electrons moving at high speed are directed at it

focal length the distance from the centre of a lens to its principal focus

force a force changes the motion of the object if it is the only force acting on the object. Force is measured in newtons (N). The resultant force on an object = its mass × its acceleration

free electrons electrons in a metal that move about freely inside the metal; free electrons have broken free from the atoms in the metal; also referred to as conduction electrons

free-fall falling at constant acceleration

frequency number of complete waves passing a point in one second, measured in hertz (Hz)

fuse a component that contains a thin wire that melts if too much current passes through it

fusion, thermal process of fusing different solids together when they melt

fusion reactor nuclear reactor (not yet developed) in which energy is released due to nuclear fusion

G

galvanometer a centre-reading meter used to show the direction of a current

gamma radiation high energy electromagnetic radiation electrons emitted by certain radioactive substances, usually after an unstable nucleus emits an α- or a β- particle; γ-radiation needs several cm of lead to stop it; it is not deflected by an electric or magnetic field and is weakly ionising

Glossary

gravitational field strength force of gravity per kilogram on an object, measured in newtons per kilogram (N/kg)

gravitational potential energy energy due to position; for an object of mass *m* which is moved up or down through height *h*, its change of gravitational potential energy = mgh

half life of an isotope the time taken for the number of nuclei of an isotope in a sample to decay by half

hard magnetic material magnetic material that is hard to magnetise and demagnetise (e.g. steel)

heat energy energy transfer from a hot object to a cold object

Hooke's law the extension of a spring is directly proportional to the weight it supports

hydraulic pressure pressure in a hydraulic system that enables a force to be exerted

inertia resistance to the change of motion of an object due to its mass

infra-red radiation electromagnetic radiation just beyond the red part of the visible spectrum; it can be detected by its heating effect

input sensor a sensor device (or a sensor circuit) that produces an electrical signal in response to a change of a physical property

input transducer a sensor circuit that provides an input signal to another circuit

internal energy energy of an object due to the motion and positions of its molecules

ionisation the process of creating charged atoms (i.e. ions); ions are created when X-rays or radiation from a radioactive source pass through a substance

isotope atoms that have the same number of protons and different numbers of neutrons in the nucleus

kinetic energy energy due to motion; for a mass *m* moving at speed *v*, its kinetic energy = $\frac{1}{2}mv^2$

latent heat of fusion energy supplied to a substance to melt it; also, energy removed from a substance to solidify or freeze it

latent heat of vaporisation energy supplied to a liquid substance to vaporise it; also, energy removed from a vapour to liquefy it

law of force between electric charges like charges repel; unlike charges attract

law of force between magnets like poles repel; unlike poles attract

law of reflection for a light ray reflected by a mirror, the angle of incidence = the angle of reflection

law of refraction for a light ray that is refracted when it travels from air into a transparent substance of refractive index *n*, $\sin i \div \sin r = n$

light dependent resistor (LDR) a resistor which has a resistance that depends on the incident light intensity; most LDRs have a resistance that decreases as the light intensity increases

light energy energy transfer by light

limit of proportionality the limit to which a spring can be stretched and still obey Hooke's Law

linearity of a thermometer the extent of the difference between the readings of a thermometer and the readings of a standard gas thermometer

liquid-in-glass thermometer thermometer that uses the expansion of a liquid as its thermometric property

live wire the wire in a mains circuit that is at a high voltage and is lethal to touch

logic gate digital circuit with an output voltage that depends on the voltage at each of its input terminals; types of logic gates include a NOT gate which has one input terminal, an AND gate, an OR gate, a NAND gate and a NOR gate, all of which have two input terminals

logic circuit a combination of logic gates

logic indicator a circuit containing an LED that indicates the logic state of an input or an output terminal of a logic circuit

longitudinal waves waves in which the direction of vibration is parallel to the direction in which the waves are travelling, for example sound waves

loudness the louder a sound, the greater the amplitude of the sound waves

loudspeaker a device that converts an electrical signal into sound waves

magnetic field the space surrounding a magnet in which a plotting compass would be affected

magnetic induction magnetism induced in an unmagnetised ferrous bar by holding a permanent magnet near it

magnetic lines of force a line in a magnetic field along which a plotting compass points

magnetic poles the ends of a magnet where the lines of force are concentrated

magnification an optical image of an object is said to be magnified if it is larger than the object itself

magnifying glass a converging lens used to view a magnified image of an object by holding the lens near the object and viewing the object through the lens

mains electricity alternating current at high voltage supplied to buildings from a distribution network

mass measure of the amount of matter in an object, measured in kilograms (kg)

mass number *A*, the number of protons and neutrons in the nucleus of an atom; the mass number of a nucleus is approximately equal to the mass of the nucleus relative to the mass of the proton

measuring cylinder instrument used for measuring volume

melting change of state from solid to liquid at the melting point

micrometer instrument used to measure lengths up to 30 mm to within 0.01 mm

microwaves electromagnetic waves between radio waves and infra-red radiation in the electromagnetic spectrum used in communications and in microwave ovens

molecule the smallest particle of a substance that can be identified with the substance

moment of force
force × perpendicular distance from the pivot measured in newton metres (N m)

momentum
this equals mass × velocity; the unit of momentum is the kilogram metre per second ($kg\,ms^{-1}$)

monochromatic light
light of a single colour

motor effect the force exerted on a current-carrying wire when it is in a magnetic field aligned at right angles (or at any non-zero angle) to the magnetic field lines

N

neutral wire the wire in a mains circuit that is earthed at the local sub-station

neutron an uncharged particle of about the same mass as the proton that is in the nucleus of an atom

newtonmeter a meter used to measure force in newtons

normal line at right angles to a boundary

nuclear energy energy released when the nucleus of an atom splits or disintegrates

nuclear fission process in which a uranium nucleus splits in two to form two smaller nuclei

nuclear fusion process in which two small nuclei fuse together to form a larger nucleus

nucleus the positively charged object at the centre of every atom which contains most of the mass of the atom and around which electrons in the atom move in orbits; the nucleus of an atom is composed of protons and neutrons

nuclide the name for different types of nuclei, each with a certain number of protons and a certain number of neutrons in its nucleus

O

Ohm's Law the current in a resistor at constant temperature is proportional to the potential difference across the resistor; the resistance does not depend on the current or the pd

ohmic conductor a conductor that obeys Ohm's Law (i.e. its resistance is constant)

optical fibre a thin transparent fibre which a light ray can travel along from one end to the other, undergoing total internal reflection if it reaches the surface

oscilloscope electrical instrument used to display electrical signals on a screen to show the variation of the signal amplitude with time; also used to measure pds and the frequency of electrical waveforms

P

parallel components components connected between the same two points in an electrical circuit; the pd across components in parallel is the same. The total current through two or more components in parallel is the sum of the current through each component

parallelogram of forces geometrical method for finding the resultant force of two or more forces

partial reflection a light ray that is partly reflected and partly refracted at a boundary

peak value the maximum (positive or negative) value of an alternating pd

pitch the higher the pitch of a sound, the greater the frequency of the sound waves

plane waves waves with straight wave fronts

polarised waves transverse waves that always vibrate along the same line at right angles to the direction of travel of the waves

Polaroid filter a filter that polarises unpolarised light

potential capacity to deliver energy

potential difference (pd) a measure of the energy transferred by a given amount of electric charge (i.e. one coulomb) when charge flows from one point to another in a circuit; it is measured in volts (V)

potential divider two or more resistors in series connected to a source of fixed pd; each resistor has a share of the fixed pd in proportion to its resistance

potentiometer a potential divider with a variable output pd

power rate of transfer of energy, measured in watts (W); power = energy transferred ÷ time taken

pressure force per unit area, measured in pascals (Pa); pressure = force ÷ area

principal focus, also focal point the point to which parallel rays of light directed straight at a converging lens are focused by the lens

principal of conservation of momentum in a closed system the total momentum before an event is equal to the total momentum after the event; momentum is conserved in any explosion or collision provided no external forces act on the objects that collide or explode.

principle of moments for any object in equilibrium, the sum of the clockwise moments about any point is equal to the sum of the anticlockwise moments about the same point

projector an optical instrument used to focus an image of an object onto a screen

proton a positively charged particle that is in the nucleus of every atom

proton number Z, the number of protons in the nucleus of an atom; all atoms of the same element have the same number of protons in the nucleus

R

radial electrical field the pattern of the lines of force surrounding a point charge (i.e. a very small charged sphere); the lines of force are straight lines that 'radiate' from the point charge

radio waves electromagnetic waves at the long wavelength end of the electromagnetic spectrum, used in communications

radioactive dating method of finding the age of a sample by measuring the proportion of unstable nuclei of a radioactive isotope that have decayed

radioactive decay the change that happens to an unstable nucleus when it emits alpha (α) or beta (β) or gamma (γ) radiation

Glossary

radioactive tracer radioactive isotope used to trace the flow of a substance through a system

radioactivity property of substances such as uranium that give out radiation all the time; the radiation is emitted because the nuclei of the atoms are unstable and become stable by emitted alpha (α) or beta (β) or gamma (γ) radiation

radiograph a photographic image taken using X-rays instead of light

random changes events that cannot be predicted

range of a thermometer the range of temperatures that can be measured from the lowest to the highest

rarefaction in a wave part of a sound wave where air molecules are spaced further apart than when they are undisturbed

real image image of an object formed by focusing light onto a screen

reflection return of waves after they reach a smooth surface

refraction change of direction of waves when they travel across a boundary where their speed changes

refractive index speed of light in air÷speed of light in a substance; symbol n

relay an electrically operated switch

resistance potential difference across a component÷the current through it, measured in ohms (Ω)

resistor an electrical component designed to have a certain resistance

resistors in parallel for two or more resistors in parallel, the combined resistance is less than the smallest individual resistance; the total resistance R of two or more resistors in parallel is given by an equation of the form:

$$\frac{1}{R} = \frac{1}{R_1} + \frac{1}{R_2}$$

resistors in series the total resistance of two or more resistors in series is equal to the sum of their separate resistances

resonance increase of the loudness of sound waves of a certain frequency that can be brought about in a musical instrument

resultant force a single force on an object that has the same effect as the forces acting on it

S

saturated vapour vapour in air that can hold no more vapour

scalar a physical quantity that has magnitude only

sensitivity of a thermometer change of the thermometric property for a 1°C rise of temperature

sensor circuit a circuit with an output pd that changes in response to a change of a physical property (e.g. temperature or light intensity)

series components components connected in the same circuit such that the same current flows through them; the total pd across two or more components in series is the sum of the pds across each component

short-circuit a fault in a circuit that provides a low-resistance path allowing a dangerously high current to flow

soft magnetic material magnetic material that is easy to magnetise and demagnetise (e.g. iron)

solar cell a cell (usually in a panel with other solar cells) that generates electricity when sunlight falls on it

solar heating panel a panel in which water flowing through it is heated by solar energy

solenoid a long coil of wire wound around a tube that may contain an iron core or be air-filled

solidifying change of state from liquid to solid at the melting point, also referred to as freezing

sound waves vibrations of layers of air that travel through the air

specific heat capacity energy needed by 1 kg of a substance to raise its temperature by 1°C

specific latent heat of fusion energy needed to melt 1 kg of a substance at its melting point; also equal to the energy that must be removed from 1 kg of a substance at its melting point to solidify it

specific latent heat of vaporisation energy needed to vaporise 1 kg of a substance at its boiling point; also equal to the energy that must be removed from 1 kg of a vapour to liquefy it at its boiling point

speed distance travelled per second, measured in metres per second (m/s)

speed of a wave, wavespeed distance travelled by a wave crest or a wave trough per second, measured in metres/second (m/s); speed of a wave = wavelength × frequency

split-ring commutator electric motor coil connections in the form of a split ring used to connect the spinning coil to the motor power supply; the split-ring arrangement ensures the current in the coil reverses every time the coil turns through a half turn so the coil continues to spin in one direction only

spring constant force/extension for a spring that obeys Hooke's Law

standard atmospheric pressure the mean pressure of the Earth's atmosphere at sea level

state of matter physical state of a substance, either solid or liquid or gas (also referred to as vapour)

static electricity electric charge on an object

sublimation change of state directly from solid to vapour

T

terminal speed speed reached by a falling object

thermal capacity energy needed by an object to raise its temperature by 1°C

thermal expansion increase of length of a solid or increase of volume of a liquid or gas due to an increase of temperature

thermal radiation electromagnetic radiation emitted by an object due to its temperature

thermionic emission process in which free electrons in a heated metal filament in a vacuum are emitted from the filament

thermistor a resistor which has a resistance that depends on temperature; most thermistors have a resistance that decreases as its temperature increases

thermocouple thermometer electrical thermometer that makes use of the voltage between two different metals in contact with each other

thermometer instrument used to measure temperature

thermometric property physical property of a thermometer that varies with temperature

total internal reflection a light ray in a transparent substance is totally internally reflected when it reaches the surface if its angle of incidence is greater than the critical angle of the substance

transformer a device consisting of a primary coil and a secondary coil wound on the same iron core; the ratio of the secondary pd to the primary pd is equal to the ratio of the number of turns of the secondary coil to the number of turns of the primary coil; a step-up transformer has more secondary turns than primary turns so the secondary pd is greater than the primary pd; a step-down transformer has fewer secondary turns than primary turns so the secondary pd is less than the primary pd used to change the peak value of an alternating pd

transformer efficiency ratio of power delivered by a transformer to the power supplied to it. For a transformer that is 100% efficient, the
primary current × primary voltage = secondary current × secondary voltage

transverse waves waves in which the direction of vibration is perpendicular to the direction in which the wave travels; example include electromagnetic waves and waves on a vibrating string or rope

truth table a table showing the state of the output voltage of a logic circuit for every combination of input states

ultrasonic waves or **ultrasound** sound waves at frequencies above 20 000 Hz, the upper frequency limit of the normal human ear

ultraviolet radiation electromagnetic radiation just beyond the violet part of the visible spectrum; it has no heating effect and is harmful to the eye

uniform electric field the pattern of the lines of force between two oppositely charged parallel plates; between the plates, the lines are straight lines directed from the positive to the negative plate

uniform wire a wire with the same diameter all along its length

unpolarised waves transverse waves that do not always vibrate along the same line

unsaturated vapour vapour in air that can hold more vapour

useful energy energy transferred for a purpose

U-tube manometer instrument used to measure the pressure of a gas relative to the atmosphere

Van de Graaff generator a machine that generates electrostatic charge on its metal dome

vaporisation change of state from liquid to vapour below or at the boiling point

vector any physical quantity that has magnitude and direction, such as force, velocity and acceleration

velocity speed in a given direction, measured in metres per second (m/s)

virtual image image of an object viewed through a lens or a mirror formed where light appears to come from

voltmeter electrical instrument used to measure emf or potential difference (i.e. voltage) in volts

wavefront points along a crest or trough of a wave as it progresses

wavelength the distance from one wave crest to the next; symbol λ (lambda); measured in metres (m)

waves disturbances that transfer energy from one place to another

weight the force of gravity on an object, measured in newtons (N)

white light light which can be split into all the colours of the spectrum, namely red, orange, yellow, green, blue, indigo and violet

work done energy transferred by a force when it moves an object or changes its shape; work done = force × distance moved in the direction of the force; work is measured in joules

X-rays electromagnetic waves that ionise substances and are produced by an X-ray tube

Index

References to tables are given in **bold** type. References to pictures are given in *italic* type.

Index